Accounting with QuickBooks®

(and QuickBooks Pro®)

with Proper Accounting

This book is designed to offer you step-by-step instruction, by using simple, layman's language and practical examples.

Gregory M. Doublas, MBA

DELMAR
THOMSON LEARNING™

Australia • Canada • Mexico • Singapore • Spain • United Kingdom • United States

**Accounting with QuickBooks and QuickBooks Pro
with Proper Accounting**
by Gregory M. Doublas, MBA

Business Unit Director:
Alar Elken

Acquisitions Editor:
Mark Huth

Editorial Assistant:
Dawn Daugherty

Executive Marketing Manager:
Maura Theriault

Channel Manager:
Mona Caron

Executive Production Manager:
Mary Ellen Black

Project Editor:
Tom Stover

Production Coordinator:
Larry Main

Cover Design:
Michael Eagan

COPYRIGHT © 2001 by Delmar,
a division of Thomson Learning, Inc.
Thomson Learning™ is a trademark
used herein under license

Printed in Canada
1 2 3 4 5 XXX 05 04 03 02 01

For more information contact
Delmar,
3 Columbia Circle, PO Box 15015,
Albany, NY 12212-5015.

Or find us on the World Wide Web
at http://www.delmar.com

ALL RIGHTS RESERVED. No part of
this work covered by the copyright
hereon may be reproduced or used
in any form or by any means—
graphic, electronic, or mechanical,
including photocopying, recording,
taping, Web distribution or information storage and retrieval systems—without written permission
of the publisher.

For permission to use material from
this text or product, contact us by
Tel   (800) 730-2214
Fax   (800) 730-2215
www.thomsonrights.com

ISBN 0-7668-3918-4

NOTICE TO THE READER

Publisher does not warrant or guarantee any of the products described herein or perform any independent analysis in connection with any of the product information contained herein. Publisher does not assume, and expressly
disclaims, any obligation to obtain and include information other than that provided to it by the manufacturer.

The reader is expressly warned to consider and adopt all safety precautions that might be indicated by the activities herein and to avoid all potential hazards. By following the instructions contained herein, the reader willingly
assumes all risks in connection with such instructions.

The Publisher makes no representation or warranties of any kind, including but not limited to, the warranties of fit-
ness for particular purpose or merchantability, nor are any such representations implied with respect to the material set forth herein, and the publisher takes no responsibility with respect to such material. The publisher shall not
be liable for any special, consequential, or exemplary damages resulting, in whole or part, from the readers' use of,
or reliance upon, this material.

About the Author

Gregory M. Doublas is an expert in computerized accounting. Greg has taught thousands of business people throughout the USA, in seminars or privately. Greg has taught people how to use accounting programs effectively and how to maximize the use of computerized accounting in their business.

Mr. Doublas holds a Masters in Business Administration (MBA) degree and is a member of the QuickBooks® Professional Advisors Program. He has developed a new practical and unique method that presents accounting in a completely new, revolutionized way and helps non-accountants understand accounting.
He has over twelve years of actual work experience in the areas of Management, Administration/Finance and Marketing in such industries as: Manufacturing, Construction, Transportation and Software Development.

In 1991 he founded Systems Management Services, Inc. for the purpose of helping educate the small and mid-size American business people in the areas of computerized accounting and management.
Mr. Doublas has over eight years of experience as a seminar presenter teaching computerized accounting and as consultant to businesses on the subjects of management and computerized accounting.
He uses common language and practical examples during his seminars and in this book.

Gregory has a unique ability to provide straightforward and simple to understand solutions to computer related business situations. He is available for consultation, either on-site or via telephone and can be contacted at gdoublas@systemmanagement.com.

Acknowledgments

I thank the Lord for the knowledge and the understanding He has provided to complete this project and my family for their patience and support.

Table Of Contents

INTRODUCTION .... **IX**

CHAPTER 1 .... **1**

STARTING IN QUICKBOOKS® 1
Using QuickBooks® *2*
Starting QuickBooks® 2
Running QuickBooks® 2
The Navigator (Versions 4, 5, 6 and 99) 2
The menu bar (Versions 4, 5, 6 and 99) 3
The iconbar (Versions 4, 5, 6 and 99) 3
The menu bar for QuickBooks® 2000 (and QuickBooks Pro®) 3
The information you need to start your company *4*
Creating your company in QuickBooks® *6*
About the Chart-of-Accounts *24*
Using the Ready-to-Use Lists (that are provided with this book) 26
The Register 27
Entering Vendors *28*
Vendor Activity Report 29
Entering Customers *30*
Customer Activity Report 31
Adding Jobs to Customer Names 32
Job Types 33
Entering Beginning Balances *34*
Items *37*
Items use list 37
Class *45*
Terms *47*
The Accounting Quickhelp Table ™ *48*

CHAPTER 2 .... **50**

PURCHASES & PAYMENTS 50
Write Checks *51*
Job Costing Expenses 54
Billable Expenses 56
Using Classes 58
Printing Checks 59

Memorizing Transactions 63
Use of Inventory 72
Editing, Voiding and Deleting 73
The Audit Trail 75
Petty Cash *76*
Transfer of Funds *78*
Purchase Orders *80*
Enter Bills *83*
Job Costing Expenses 85
Job Costing Expenses 86
Converting P.O's into Bills 87
Closing a Purchase Order 90
A/P report 91
Credit Card Charges & Credits *92*
Job Costing Expenses 95
Reconciling the Credit Card Statement 97
Pay Bills *102*
Purchase Discounts 102
Using the Register *105*
Pay Sales Tax *108*
Sales Tax Credit 109
Form 1099 .... *110*

CHAPTER 3 112

INCOME & CUSTOMER PAYMENTS 112
Create Estimates *113*
Customer Prepayments *116*
Make Deposits *119*
Custom Reports 121
Keeping Cash Back 122
Deposit Report 123
Create Invoices *124*
Job Costing Income 124
Departmentalized Accounting 124
The A/R Report 126
Converting Estimates to Invoices 126
Progress Billing 128
Receive Payments 130
Write-off Bad Debt 133

Create Credit Memos/Refunds	*134*
Issue Credits to Customers	135
Issue Refunds to Customers	135
Enter Cash Sales	*136*
Enter Statement Charges	*140*
Assess Finance Charges	*141*
Create Statements	*142*
Customer Bad Checks	*143*

CHAPTER 4 .... **147**

PAYROLL	147
Payroll Items	*148*
Payroll Items use list	148
Enter Employees	*165*
Tracking Employee Time	*168*
Time Reports	171
Invoicing Billable Time	172
Pay Employees	173
Print Paychecks	175
Edit Paychecks	175
Payroll Reports	176
Pay the Liabilities and Taxes	176
Payroll Forms	*178*
941 Form	178
940 Form	179
W-2 Form	181

CHAPTER 5 .... **182**

OTHER FUNCTIONS & MISC. INFORMATION	182
Make Journal Entries	*183*
Bank Account Reconciliation	*188*
Write Letters	*192*
Accountant's Review	*194*
Timer Activities	*197*
Invoicing Billable Time	200
Export Addresses	*201*
Exporting Files	*203*
Backup & Restore	*204*

vii

Customizing Forms	*205*
Budgeting	*212*
Year-end Process	*215*
QuickBooks® Keyboard Shortcuts	*216*
Glossary	*217*

CHAPTER 6 .... **219**

REPORTING	219
Profit & Loss Statement	*220*
Balance Sheet	*221*
Trial Balance	*223*
A/P Report	*223*
A/R Report	*224*
Deposit Report	*224*
Job Profitability Report	*225*
Getting the Lists	*225*
INDEX	*227*

Introduction

This book was written for you, the QuickBooks® user, with the main purpose of helping you get the most out of computers and the software.

As you already know, the QuickBooks® software program is the most popular accounting program on the market. It can track all of your company's business transactions. It is an excellent business tool that when used properly, will provide you with the information you need to manage your company more effectively.

This book will help you learn not only how to use the QuickBooks® software program, but to gain an in-depth understanding of the accounting you're completing each time you record a business transaction.

You will also learn to use the QuickBooks® software to its full potential. Through practical examples, we'll show you how to utilize powerful functions that will allow you, for example; to track Income & Expenses for Job Costing and Departmentalized accounting purposes so that you can maximize the value of the QuickBooks® software to your company.

This book will provide the information you need on a step-by-step basis on such subjects as:
1. How to record all the *regular* transactions in the QuickBooks® software such as invoices bills and payments. How to do the payroll and pay the payroll taxes and etc.
2. How to record *special* transactions such as *prepayments, bad debt, refunds* and etc.
3. How to use *proper accounting* for each transaction so that your reports will be accurate.
4. How to use key functions in the QuickBooks® software such as the Import, Export, Accountant's Review, and Timer Activities. How to Backup & Restore data and much, much more.
5. How to use the P/L and the Balance Sheet and other important reports for good decision-making.
6. And in case you are new to the QuickBooks® software program, we'll show you how to set-up your company and enter the beginning information.

In this book, we have also included a section we call *business tips* to help you follow proper business procedures and a handy *glossary* that explains important computer and business terms. Also, we've included our Accounting Quickhelp Table™ to help you understand the various transactions you need to record on a daily basis.

Thank you for purchasing my book and I wish you the best of success in your effort to learn how to use the QuickBooks® program and accounting the proper way.

Gregory M. Doublas, MBA

Chapter 1

Starting in QuickBooks®

The purpose of this chapter it to help you learn how to start your company in the QuickBooks® software program, create new accounts in the chart-of-accounts, enter new vendors, customers, employees and all the important information you'll need to make computerization meaningful.

In this chapter we will be discussing the following subjects:

- **Using QuickBooks®**
- **The information you need to start your company**
- **Creating your company in QuickBooks®**
- **Using the Chart-of-Accounts**
- **Entering Beginning Balances**
- **Entering Vendors**
- **Entering Customers**
- **Items**
- **Class**
- **Terms**
- **The Accounting Quickhelp Table©**

Using QuickBooks®

Starting QuickBooks®

There are two ways to start the QuickBooks® program: By using the **Start** button or via a **Shortcut** icon on the Windows® desktop

1. To start QuickBooks® through the **Start** button, click at:
    a. The **Start** button, on the lower left corner on the Windows® desktop
    b. Next, click at the **Programs** option from the menu
    c. Select and click once on the QuickBooks® **icon** from the next menu

2. Using a **Shortcut** icon:
    If you have a **Shortcut** on the Windows® desktop, simply click *twice* on the Shortcut icon and QuickBooks® will start.

Running QuickBooks®

Once you have started QuickBooks®, you may run it by choosing one of the three different options available to you or by using a combination of all three at various times. The three options available are the following: The **Navigator**, the **menu bar,** and the **iconbar**.

No matter which way you choose, the result of using QuickBooks® will be the same, whether you use one of the three options or a combination.

The Navigator (Versions 4, 5, 6 and 99)

The **Navigator**, as its shown in figure 1.1, is a way to find lists, forms, the registers and the reports in QuickBooks®. You may also activate the various functions available to you via the Navigator by clicking on a particular *icon* or name.

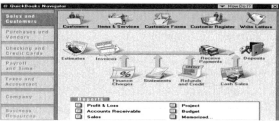

Figure 1.1 shows the **Navigator**.

The menu bar *(Versions 4, 5, 6 and 99)*

The **menu bar** contains all the QuickBooks® functions such as, lists, reports and forms you may need to use on a daily basis. The menu items you will most likely use most will be the following:

1. The **Activities** menu item contains all the main functions you need to record all your business transactions such as: to bill customers, record purchases and payments, do payroll, pay taxes and etc.
2. The **Lists** at the menu bar, contains all the lists such as: The Chart-of-Accounts, Customer, Vendor, Employees, Items, Payroll Items and etc.
3. The **Edit** menu item contains auxiliary functions such as The Memorize, Delete, Void, Find, Change Account Color and etc.
4. The **File** menu item allows access to files for such purposes as to: customize QuickBooks®; to change the name, address, city, state, zip, fiscal year, etc. to your company's file; to print forms such as checks, invoices, paychecks, purchase orders, etc.
5. The **Reports** menu item contains most of the reports you will need to print on a regular basis such as: Weekly, monthly, or at other times.

File Edit Lists Activities Reports Online Window Help

Figure 1.2 shows the **menu bar**.

The iconbar *(Versions 4, 5, 6 and 99)*

The iconbar is another way to run QuickBooks®. It contains pictures or buttons and its purpose is to allow you limited access to QuickBooks® functions that are represented by the icons.
To use a function via the iconbar, all you have to do is *click* once on an icon or button.

Calc Estimate Invoice PO Time Check Bill Accnt Cust Vend Item Find Help

Figure 1.3 shows the **iconbar**.

The menu bar for QuickBooks® 2000 *(and QuickBooks Pro®)*

The menu bar for QuickBooks® 2000 allows complete access to all program functions.

File Edit Lists Company Customers Vendors Employees Banking Reports Window Help

Figure 1.2a shows the **menu bar** for QuickBooks® 2000

1. The **File** menu item allows access to files for such purposes as to customize the company's name, address, city, state, zip, fiscal year, etc. To print forms; to Backup and Restore; provides access to the Utilities the Timer and the Accountant's Review functions.
2. The **Edit** menu item provides access to such functions such as the Memorize, Delete, Void, Find, the Preferences, the Register and etc.
3. The **Lists** menu item contains all the lists available such as: the Chart-of-Accounts, Customer, Vendor, Employees, Items, Payroll Items, and access to the Templates and etc.
4. The **Company** menu item provides access to information about the company, QuickBooks® services, etc.
5. The **Customers** menu item provides access to things that pertain to your **customers** such as information about customer activity; the Invoice, Estimate and other customer oriented functions and the Time Tracking.
6. The **Vendors** menu item provides access to things that pertain to your **vendors** such as information about vendor activity; the Enter Bills, Pay Bills and other vendor oriented functions and Inventory Activities functions.
7. The **Employees** menu item provides access to things that pertain to your **employees** such the Pay Employees and Pay Payroll Liabilities; the Payroll Forms and W-2s; Time Tracking and other important functions.
8. The **Banking** menu item provides access to creating disbursement Checks and banking related activities such as Make Deposits, Transfer of Funds, Credit Card Charges, the Reconcile and other important functions.
9. The **Reports** menu item contains reports such as the P&L and the Balance Sheet that you will need to monitor your company's financial and operating performance. Customer and Vendor related information and other important business reports.

The information you need to start your company

$\mathcal{T}$o start your company properly in QuickBooks®, you need to have the right kind of information about your company entered into the QuickBooks® program. There is an ancient Greek proverb that says: *The begin is half of everything you try to accomplish*. It means that the beginning is extremely important. And so is the case in computerized accounting.

A proper start involves two issues: **Time** and **information**.
1. **Time** is about *when* to start your company? Certain times during the fiscal year will determine the *quantity* you may have to deal with, the *quality* and the *cost* of entering the information. The further away you are from the end of the fiscal year the more the information you may have to work with. That means, the more time it will take you to enter your company information. And of course, time is money whether you do it on your own or hire the services of your accountant.
Likewise, a large quantity of data will possibly have a negative affect on the accuracy or quality of your data in the system.

Here is a hierarchy of the *best* to the *least best* of possible times you can start a new company:
    a. You may start at the end of a *fiscal* year.
    b. You may start at the end of a *quarter*.
    c. You may start at the end of a *month*.

2. To enter the information properly, you must understand two important dates:
The **start** and the **current** date (the date you are now). For example: If you're on Feb. 20 now, that's your *current* date, you may want to start as of Jan.1. That's the *start* date.
The information you'll need to have is the following:
   a. Your company's *Federal ID* number.
   b. Know the first month of your *tax* year and the first month of your *fiscal* year.
   c. Have a *Trial Balance* (please consult with your accountant about the Trial Balance).
   d. Have available Customer, Vendor and Employee *name,* and *addresses*.
   e. Have all the *open bills* your company owes to *vendors* as of the current date.
   f. Have all *open customer invoices (money* owed to your company) as of the current date.
   g. Have available all paid *purchases* by check number from the start to the current date.
   h. Have available all paid *invoices* issued to customers from the start to the current date.
   i. Have available all the *checks* written from the start to the current date.
   j. Have the *Payroll* total figures accumulated by <u>employee</u> and on a <u>quarterly</u> basis.

Creating your company in QuickBooks®

Creating a new company in QuickBooks® is a simple step-by-step process you can

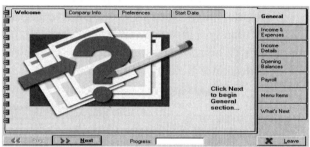

follow easily. This process is called the **Easy Step Interview** as shown in figure 1.4. Figure 1.4 shows the **Easy Step Interview** process.

To create a **new company**, follow the steps below:
1. Select **File** at the menu bar and the **New Company** option from the drop down menu.
2. At the **Easy Step Interview**, select the **Next** button to proceed to the next step.
3. Allow the default position of "**No I'm not Upgrading**" to stay
4. Continue selecting by pressing the **Next** button until you come to the **Your company name** screen. Enter your Company's name, Address, City, State, and Zip code.
5. Next, press the **Next** button and enter your company's federal ID number. This is a number issued by the Federal Government and looks like 35-1234567.
6. Select from the arrow button below, the first month of your company's *tax* year. The business tax year is a twelve-month period you choose to report the company's income taxes. Please consult with your accountant in order to select the proper month. In the field below, select the arrow button the company's first month of the *fiscal* year. The fiscal year is a twelve-month period you choose to track the company's business activity.
7. On the next screen, you need to select a tax form that is appropriate to your company. If you are a regular Corporation, select *Form 1120*. If you are an S Corporation, select *Form 1120S*. For a partnership, select *Form 1065*. *Please consult your accountant before you select the tax form of your company.*
   The selection you make here will be used to connect your data in QuickBooks® with the TurboTax® program. If you do not intend to use TurboTax®, select the "Other/None" option.

8. Next, select the *Industry* that you feel best *matches* your company. The selection you make here will determine the type of *chart-of-accounts* that will be selected for your company.
9. Next, in the *Save As* screen, allow QuickBooks® to proceed with the default name (this is the name you have typed during the previous step for your company). Using this name, QuickBooks® will create your company file. This is the name that becomes the company's directory or "ID". Next, click at the **Save** button and your company's name will be saved in the QuickBooks® folder. Please notice the company's name at the **File name** field as its shown in figure 1.

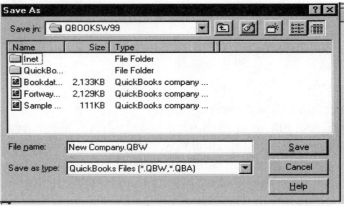

Figure 1.5 shows the **Save As** screen

10. In the "*Your Income and expense accounts*" screen you will be dealing with the accounts in the chart-of-accounts. You can view the chart-of-account list that appears on this screen by scrolling up or down (point with the mouse button on the upper or lower arrow buttons).

If you like these accounts, you may select the **Yes** option to the left to retain them, or the **No** option to *create your own*. If you select the Yes option, you still can add or edit the existing accounts (or even delete the ones you don't need as long as you do it before you add any data).

If you select the **No** option, you may add accounts on your own at a later time or you may elect to use the accounts that are provided with the purchase of this book. To use the chart-of-accounts (that come with the book), simply follow the instructions on page # 226.

Adding new accounts to the chart-of-accounts will be addressed in the **About The Chart-of-Accounts** section below.

11. The *"How many people"* screen allows you to tell QuickBooks® how many people will have access to your company data. Enter the number of people that may use your company data in QuickBooks®. When you finish the Interview process, QuickBooks® will remind you to enter the User names and their passwords. It's an optional selection.
12. Next, after you click three times on the **Next** button below, you will be in the **Preferences MODE**. The Preferences allow you to customize the way QuickBooks® will work for you. Most of the selections are self explanatory except the following ones which we will explain here:
    a. In the **Progress Invoicing** screen, select the **Yes** option for *partial* customer billing if you plan to invoice customers in increments instead of converting the entire estimate in full.
    b. In the **Time tracking** screen select the **Yes** option to keep track of time (this is how you will track your billing for the customer).
    c. Next, select **Yes** to use *Classes*. They play an important role, which we will explain later.
    d. Next, **do** select the *"Enter the bills first and then enter the payments later"*.
    e. Next, select the *"At start up"* option to start the Reminders when you start QuickBooks®.
    f. Next, select the *"Accrual-based reports"* option. For an explanation of cash vs. accrual base accounting see the Glossary section. I recommend that you select the Accrual base accounting because it will allow you to have meaningful financial statements that would help you manage your company better.
    In the event that you need to have the information for tax purposes to be on the Cash basis, QuickBooks has the capability to allow you to print it on the Cash basis, even though you will be doing accounting that is based on the Accrual base.
13. After two clicks, you are in the *Start Date* option. Select the date you will start your company in QuickBooks®. This is the date that you will be adding the information about your company for the first time in QuickBooks®.
    Your start date must be the beginning of a quarter. Here are some examples:
        a. If you are now on Feb. 10, you want to select Jan. 1 as a start date.
        b. If you are now on April 20, you want to select March 31 as a start date.
        c. If you are now on July 10, you want to select June 30, as a start date.
    We'll talk next on how to enter the beginning information.

14. The tab to the right, *"Income & Expenses"*, allows you to enter Income and Expense type accounts in the chart-of-accounts.

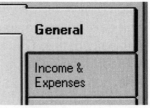

Figure 1.6 shows the **Income & Expense** tab that is used to enter new accounts.

You may enter new accounts through this function; through the **Lists** at the menu bar and the **Chart-of-Accounts** option, or you may use the chart-of-accounts we have for you by downloading from our website (see instructions on page # 226).

To enter an *Income* account, select the **Yes** option and type the *name* of the account you want to enter such as *Sales Product A* and click the **Next** button. Repeat the process for another account.

To enter an *Expense* account, select the **Yes** option and type the name of the account you want to enter such as Office Supplies and click the **Next** button. Repeat the process for another account.

15. The tab to the right, *Income Detail*, allows you to enter Items such as Service, Inventory Part, Non-Inventory and Other Charges type Items. In the **Items** section later, we'll discuss the purpose and how to enter new Items.

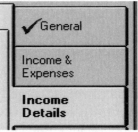

Figure 1.7 shows the Income Detail tab.

16. The *Opening Balances* tab allows you to enter **Opening Balances**.

9

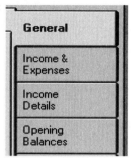

Figure 1.8 shows the *Opening Balances* tab.

Opening Balances are amounts of money your company owes to other companies (or people) and the amounts of money customers owe to your company.
Opening Balances can be entered for:
- Customers
- Vendors
- Credit cards
- Bank loans

To enter **Opening Balances** for each type, follow the steps below:
a. To enter *customer* open balances, select the **Yes** option at the *Enter Customer* screen and **Yes** again if you want to track the amount into a Job or Project (for Job Costing as we will see later). Next, type the Job name, the *amount* owed and click at the **Next** button to finish.
The Opening Balances that you enter through this function will be in a summary format. If you want detail, with each invoice showing separately, you may want to enter opening balances the way it will be described in the *Enter Beginning Balances* section.
b. To enter *vendor* open balances, select the **Yes** option at the *Adding Vendor with Open Balances* screen. Next, type the vendor name and the *amount* owed, and finishes by clicking at the **Next** button.
The Opening Balances that you enter through this function will be in a summary format. If you want detail, each amount showing separately, you may want to enter opening balances the way it will be described in the *Enter Beginning Balances* section.

c. To enter *Credit Card* balances, select the **Yes** option and at the *Credit Card Accounts* screen and type the name of the credit card. Next, set the *Ending date of the statement* at the next screen and type the *amount* owed to the credit card. Click at the **Next** button to finish.
d. To enter a *Line of Credit* open balance (an amount owed to the Bank) select the **Yes** option and type the name of the loan like *State Bank X* on the next screen. Select the **Next** button below and type the *Statement Ending Date* and the *amount* owed. Click at the **Next** button to finish.
e. To enter a *Loan or Note* payable balance, select the **Yes** option and at the next screen type the *name* of the loan, such as *Loan State Bank X*. Click at the **Next** button and type the *Statement Ending Date* and the *amount* owed. Click at the **Next** button to finish.

17. The *Payroll* tab allows you to enter information that plays a key role in the performance of the payroll in QuickBooks®.

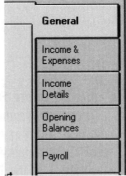

Figure 1.9 shows the *Payroll* tab.

The payroll tab allows you to customize QuickBooks® to fit your company's needs. You may enter such information as Federal and State ID's, FUTA and SUTA rates, etc.

Through this part of the Interview, you will also be creating *Payroll Items* that will help you further customize QuickBooks® (Payroll Items can also be created via *Lists* at the menu bar and the *Payroll Items* option) and thus be able to do payroll properly.

a. The *Pay period* screen defines the frequency such as Weekly, Monthly, etc., of your payroll. Select one and proceed to the next option by pressing the **Next** button below.

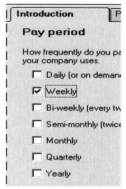

Figure 1.10 shows the **Pay period** screen.

b. *Filling states* screen allows you to select the State (s) you must withhold taxes for. You can have multiple selections.
c. In the *State Withholding Tax* screen, type in the *State Employer ID* under the State Employer ID field.
d. In the ***State Unemployment Tax*** screen, enter the *State Employer ID* and the *rate* (which is the Experience Rate) in the *ER Rate* field (stands for Employer Rate) that your State has determined for your company.
This rate is provided by the State in writing at the beginning of every calendar year. It is used to calculate the State Unemployment tax which must be paid every quarter and before the end of the following month. The *EE Rate* field (stands for Employee Rate) is for an employee paid rate that may apply in some states.
e. At the ***State Disability Tax*** screen, enter again the *State employer ID*.
f. In the ***Federal Unemployment*** or FUTA tax screen, you may select the lower rate of 0.8%. *If you are not sure which rate is applicable to your company, please consult with your accountant.*
g. In the *Compensation* screen, make the selections that apply to your type of pay. This selection will divide for example, salary pay into <u>weekly</u> or <u>monthly</u> increments and etc.

h. The *Select your hourly wage types* allows you to select Payroll Items that are Hourly type that you would be using later to pay hourly employees via the payroll, by placing a checkmark (√).

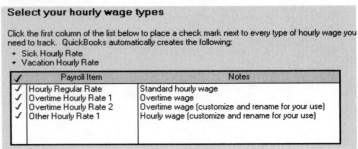

Figure 1.11 shows Hourly wage items.

Note: You may select up to the maximum of eight **Hourly** type items per employee.

The multiple items will allow you to pay different types such as regular, overtime 1, overtime 2, etc. The Items you create can be used for different employees.

i. In the next screen, *Select your salary wage types,* you need to select, by placing a checkmark (√), **Salary** type items that you may use to pay Salaried employees.

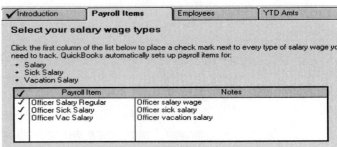

Figure 1.12 shows Salary wage items.

j. You can make multiple selections to cover various types of salaried employees such as the Bookkeeper, Officer, etc.
k. Next, in the ***Select your deductions*** screen, you need to select, by placing a checkmark (√), **Deduction** type Payroll Items.

13

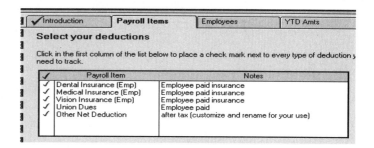

Figure 1.13 shows the **Deduction** Items.

These types of Payroll Items, are used for **deducting** from employee paychecks for such purposes as union dues, child support, garnishments, employee advances, group health insurance, etc.

l. In the ***Select your payroll additions*** screen, you need to select, by placing a checkmark (√), **Addition** type Payroll Items.

Figure 1.14 shows Additions Items.

Addition items are used to **add** to paychecks for such purposes as bonuses, reimbursement for mileage, incentive type pays and etc.

m. The ***Select your company contributions*** screen, allows you to select Payroll Items that are designed to track payroll expense that are paid by the company on behalf of the employee. For example, a *company-paid* health insurance can be tracked using a Company Contribution Payroll Item.

n. The ***Earnings*** screen allows you to select Wage type Payroll Items that will be the default Items that you may use later in the **New Employee** screen when you will be adding new employees. You may select (by clicking under the checkmark column) multiple Items.

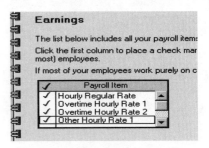

Figure 1.15 shows the **Earning** screen.

o. In the ***Standard payroll items*** screen, you select Addition, Deduction & Company Contribution type Payroll Items that will be default Items and be used for the same purpose as described in step **m** above.

p. At the ***Federal taxes*** screen make the selections that apply to your company.

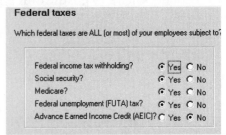

Figure 1.16 shows the **Federal** tax screen.

q. At the *State taxes* screen, select the **State** that you will file the payroll taxes.

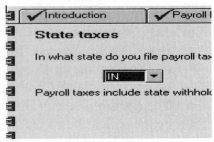

Figure 1.17 shows the **State** taxes field.

r. In the ***Sick time*** screen, select the **Yes** option if your company pays employee sick time. At the next screen, select the *Every pay* period option if you intend to have QuickBooks® accrue sick pay <u>every time</u> you do payroll. In the *How many hours...* field below, type the amount of hours you want employees to receive every time you will be doing payroll i.e. **.8** hours, rounded-off, (if the MAX. sick pay was 40 hours per year based on the 52 weeks per year, then 40 / 52 =. 769). In the field below, type the max. Of sick hours, i.e. **40**.
By clicking at the last field, it allows you to tell QuickBooks® to **reset** the accrued sick hours each year. If you do not click, the sick hours will not reset each year.

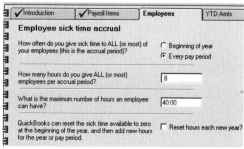

Figure 1.18 shows the *sick* time accrual screen.

s. To complete the ***Vacation time*** screen next, follow the steps outlined in step **q**.

t. The ***Adding an employee*** screen allows you to add new employees. You may also add new employees through **Lists** at the menu bar and the **Employees** option. The process of adding a new employee is the same whether you add an employee through this screen or via the *Lists/Employee* option.

To **add** an employee, follow these steps:
- At the *Adding an employee* screen, select the **Add Employee** button and click once on it.

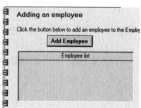

Figure 1.19 shows the **Add** employee button.

16

- In the *Address Info* tab, type a salutation such as Mr., Ms., etc., next fill in by typing the First, Middle, and the Last names. Also, the Address, the City, State, Zip code and SS Number.

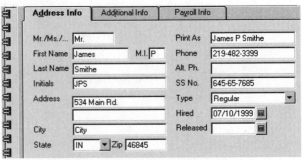

Figure 1.20 shows the **Employee** information screen.

- Next, select the **Payroll Info** tab, by clicking once and enter the following:
- In the *Earnings* field, type an hourly rate next to an *Hourly* Wage item and an annual amount next to a *Salary* Wage type item as shown in figure 1.21 below.
- In the *Additions, Deductions and Company Contributions* field below, type an amount next to the appropriate Item. It is here that you use the various Payroll Items such as Deductions, Additions, ETC. by connecting them to an employee in order to affect that employee's paycheck.
- In the *Class* field, select a class that can be used to track the employee's wage into a Class report for such purposes as Workman's Compensation. To add a new Class, click at the button next to it and then select the *<Add New>* option. At the *Class Name* field, type a name such as Office Employee and click the **OK** button. You have finished creating a Class.

Classes will track the employee wage by job classification such as Office, Field, Shop, Driver, etc., that you can use later in the event that the Workman's Compensation auditor will audit you.

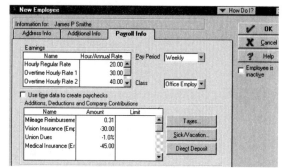

Figure 1.21 shows the employee *earnings*, *additions* and *deductions* screen
- Next click once at the **Taxes** button as shown in figure 1.21.
  - In the **Federal** tab, select the employee's *Filing Status*, the number of *Allowances*, and if any, type an amount for extra withholding the employee may designate, in *the Extra Withholdings* field (this field is capable of managing withholdings that are in addition to the Federal Tax tables).

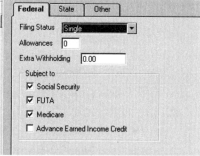

Figure 1.22 shows the *Federal* tax tab.
- Next, click once on the **State** tab and make the appropriate selections about the State the employee *Worked*, *Lived,* and the *Filing Status* as shown in figure 1.23.

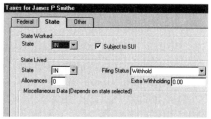

Figure 1.23 above, shows the *State* tax tab.

18

- The **Other** tab is to allow you to define a local tax withholding such as City or County tax.
  At the **Name** field, click once at the button and select the *<Add New>* option. Next, select the State from the *User-Defined Tax* option button and choose whether the tax is *Employee* or *Company* paid by clicking at the appropriate circle.
  Next proceed by typing the *name* of the agency such as State Dep. of Revenue that will receive the tax. Select next the *liability* account (an account that is *Other Current Liability* type) where the tax will accumulate until you print the check as its shown in figure 1.24.

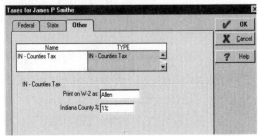

Figure 1.24 shows the *Other* tax tab.

- In the *Print on W-2 as* field type the **name** of the locality. This name will print on the W-2 form. Next, type the percent that you must withhold i.e. 1%. Click at the **OK** button.
- Next click once on the **Sick/Vacation** button. The information entered during the creation of the Sick and Vacation Payroll Items at the **q** and **r** steps above, should appear here in the *Sick & Vacation* screen. To save your selections, simply click the **OK** button. Once again click the **OK** button at the *New Employee* screen to save the information you have entered for the new employee.

To add another employee, click once at the **Add Employee** button and repeat the process.

Entering Year-to-date (YTD) Payroll Information:

The *Year-to-date (YTD) Amounts* screen allows you to enter summaries of employee paycheck information for prior months. This information should be entered by calendar quarter.

The payroll information you enter through this function, will result in the following:
1. Accurate payroll reports on liability and expense accounts
2. Accurate employee payroll summaries
3. Accurate W-2 information

YTD payroll information can also be entered, with the same results, by selecting the following options:
1. The **Activities** at the menu bar
2. The **Payroll** option from the drop down menu
3. The **Set Up YTD Amounts** option.

To enter **YTD** payroll information:
1. At the **Payroll: Year-to-date (YTD) Amounts** screen, press the **Next** button to proceed to the next screen.
2. In this new screen, type or select the **date** that you want your accounts in the chart-of-accounts, such as the liability and expense accounts, to be *affected* by the year-end payroll amounts you will be entering. This date should be a quarter-end date such as 3/31 or 6/30, etc., then click at the **Next** button to proceed to the next screen.
3. Next, enter the *date* you will begin using the Payroll function to create paychecks.
4. At the **Employee YTD summaries** screen, as shown in figure 1.25 below, highlight an employee's name and click the **Enter Summary** button.

Figure 1.25 shows the **Employee YTD summaries** screen.

In this screen, select the following:
a. The checking account at the **Bank Account** field at the top type.
b. Check the *Cleared* field and type an alphanumeric "number" at the **No** field.
c. Set the *date* as of the *last quarter* at the **Date** field,
d. In the **Salary and Hourly Wages** field below, type the payroll *amount* under the **Period Amount** column and the *hours* worked under the **Hours for Period** column.
e. In the **Other Employee and Company Payroll Items** field, type in the amounts you *paid* for mileage reimbursement, bonuses, etc. and the amounts you *deducted* for such things as: child support, garnishments, union dues, health insurance, etc., and the amounts you *withheld* for the Federal, State and Local governments.
f. Next, at the *From* and *To* fields, select the dates for the historical paycheck.
g. Next, click at the **Accounts Affected** button and select one of the three options. The selection that you make here is very important because you can affect your accounts in the chart-of-accounts.
If you are beginning now in QuickBooks®, you <u>do</u> want to select the **Affect liability, expense, and bank accounts** option. This selection will affect the *liability* and *expense* accounts that are related to payroll via the payroll items and the transactions will be recorded in your system.
h. Click at the **OK** button to save your entries.

Figure 1.26 shows **YTD** adjustments to the payroll.

21

To enter YTD information for another quarter or month for the same employee, click at the **Next Period** button, on the upper right corner.

Recording Payments for Taxes and Liabilities:

The amounts of the deductions and the withholdings (liabilities) we have entered above, must now be *adjusted* because the data we've entered in the previous step, were from prior periods (the past) and by now, most likely, you have already paid these amounts.
The adjustment to these amounts is accomplished by recording the <u>actual</u> payments made for such things as child support, union dues, Federal, State, Local taxes, etc.

Prior payment of deductions and withholdings can be recorded also by selecting:
1. The **Activities** at the menu bar
2. The **Payroll** from the drop down menu
3. The **Pay Liabilities/Taxes** option.

To record **prior payments** of withholdings and deductions through the **Interview** process:
1. Select the **Next** button to proceed to the next step.
2. At the **Summarizing prior liability payments** screen, click the **Create** button as shown in figure 1.27.

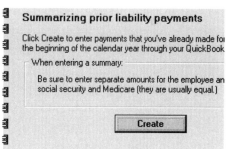

Figure 1.27 shows the **Summarizing prior liability payments** screen.

3. Next, select the *date* the payment was made at the **Payment Date** field and the *period* the payments was for at the **Period Ending** field.
4. Next, place your cursor under the **Item Name** field and click once. From the drop down menu that appears, select a Payroll Item. Next to it, at the **Amount**

field, type the amount paid. Continue until you have entered all required amounts.
5. Next, click on the **Accounts Affected** button, to the right, to designate whether you want the accounts in the chart-of-accounts to be affected. If you are starting now in QuickBooks®, you do want to affect the accounts by selecting the **Affect liabilities and bank account** option. Next, select a bank account at the **Bank Account** field and you have finished recording the payment of withholding taxes and deductions.
6. Next, click at the **OK** button to record your entries. Click at the **done** button to close this function.

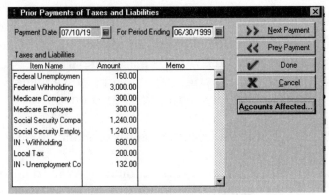

Figure 1.28 shows the *Prior* Payments of Deductions and Liabilities screen.

About the Chart-of-Accounts

*T*he Chart-of-Accounts (COA) is the heart of your accounting program. All the business transactions are recorded through the accounts in the COA.
As you record the amounts that pertain to a business transaction, the accounts you select drive the amounts into the Journal and eventually into the *financial statements* such as the Balance Sheet and the Profit & Loss (P/L).

The accounts in the COA are grouped into five categories. The five financial categories are the following:
1. Assets
2. Liabilities
3. Equity
4. Income (or Revenue)
5. Expenses

The *Asset* category has such accounts as the Bank Accounts (Checking accounts), Accounts Receivable (A/R), Petty Cash, Employee Advances, Fixed Asset accounts (Land, Building, Equipment) and etc.

The *Liabilities* category has such accounts as the Accounts Payable (A/P), Credit cards, Bank loans, Sales Tax Payable, the Payroll Withholdings, and Deductions accounts and etc.

In the *Equity* category, there are such accounts as Capital Stock, Retained Earnings and the Owner's draw account.

In the *Income* category, there are accounts that allow you to record the *sales* of the product or services your company provides to customers.

In the *Expense* category, there are such accounts as Wages expense, Payroll Tax Expense, Rent, Material Purchases, Marketing and Advertising, Telephone, Utilities, Interest Expense, Office Supplies, Maintenance, Auto Expense, and etc.

Please note: For a thorough understanding of the financial **categories**, the **accounts** in the chart-of-accounts, and how the process of **accounting** works, we highly recommend the "Understanding

Accounting" Videotape developed by the author of this Book. In the tape, the author uses practical examples that he presents in a simple, layman's language. To order, call 219-482-3399, Fax 219-471-5302.

You may create new accounts any time there is a need for one.
To create a **new** account, select the following:
1. The **Lists** at the menu bar
2. The **Chart of Accounts** from the drop down menu

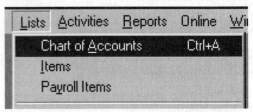

Figure 1.29 shows the *location* of the chart-of-accounts.

3. Next, select the **Account** button, and the **New** option.
4. Next, you need to select the *type* of the account by clicking next to the **Type** field and selecting from the drop-down menu. The *Type* of the account is very important. It determines the *posting* or the *position* each account will have on the financial statements.
5. Next, type a *name* for the new account at the **Name** button.
6. Next, click the **OK** button to save the new account in the system. You have just created a new account.

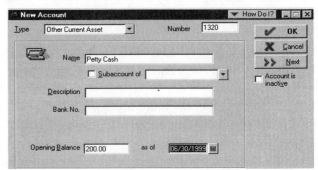

Figure 1.30 shows the **New Account** screen.

When you create a new account in the chart-of-accounts, you have the option of entering an *Opening Balance* as shown in figure 1.30. The Opening Balance is an amount (a

balance) the account may bear at the time you create it. This is the case of existing accounts. Account already in use when you start your company in QuickBooks®.
Also, you can enter *beginning* account balances via journal entries through the **Make Journal Entry** function. We will discuss how to enter beginning balances via the journal entries at the *Enter Beginning Balances* section.

When you create a new account in the chart-of-accounts, you have the option to assign a number to it. Many users and particularly accountants prefer to have their accounts associated with numbers. Numbers are typed in the **Number** field, at the upper right corner of the screen as it shows in figure 1.30 above.
Numbering the accounts in QuickBooks® is not mandatory but optional.

To have the capability to assign **numbers** to your accounts, you must first setup the **Preferences** in QuickBooks®.
To activate the capability to number your accounts in the COA, select the following options:
1. The **File** menu bar item
2. The **Preferences** at the drop down menu
3. In the Preferences, click on the **Accounting** option
4. Select the **Company Preferences** tab
5. Next, select the **Use account numbers** option

After you make the selections in the Preferences, any time you attempt to create a new account in the chart-of-accounts, on the upper right corner of the screen, there will be a new field called **Number**. That's where you may type a number.

Note: To save you time, we are providing two Chart-of-Accounts, Items and Payroll Items that you can use in your own company (please, refer to the last page for instructions on how to get the lists). To use these lists, simply follow the instructions below.

Using the Ready-to-Use Lists (that are provided with this book)

𝒯he **Chart-of-Accounts**, the **Items,** and the **Payroll Items** lists that came with this book are ready to be used. You may want to change the names of the Items and the Accounts in order that they fit your particular company needs or you can create additional new

Items and Accounts, as you need them. Using the ready-to-use lists will save you a great deal of set-up time especially if you are starting a brand new company.

To use the **Lists**, follow the example below:
For example to use the *Chart-of-Accounts* list,
1. **Download** the Lists from our website by following the instructions at the end of this book (page # 226) on **Getting the Lists**.
2. Next, in QuickBooks® select **File** (in QuickBooks® 2000 select **File / Utilities**).
3. Select the **Import** option from the side menu.
4. At the "**Look in**" field, select the drive **(C:)**, and the **folder** you've created (Download).
5. From the icons that appear in the white screen, select the chart-of-accounts icon (or any other icon, for another list) by clicking on it *once*.
6. Next, click the **Open** button below and you have just managed to import the *Chart-of-Accounts* list in your company file.

To use the *Items* or the *Payroll Items* lists, simply follow the same steps that were used for the C*hart-of-Accounts*.

The Register

*T*he **Register** is part of the chart-of-accounts and it contains all the activity that affects an account's balance. Registers exist for Asset, Liability and Equity type accounts.

Through the **Register** of an account, you may record many transactions that affect the account such as entering bills, checks, bank fees, interest income, and others.

To open the **Register** of an account, select the following:
1. The **Lists** at the menu bar
2. The **Chart of Accounts** option from the drop down menu
3. In the chart-of-accounts, *highlight* the account you want to work with and *click* twice on it. Another way to access the **Register** of an account, is after you highlight the account, select:
   a. The **Activities** button at the bottom of the screen and
   b. The Use Register option.

Entering Vendors

Vendors (or suppliers) are companies or people from whom your company *purchases* product or services.

To create a *new vendor*, select the following:
1. The **Lists** at the menu bar and
2. The **Vendors** option at the drop down menu as shown in figure 1.31.

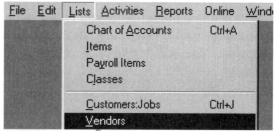

Figure 1.31 shows the location of the *Vendors* option (in QuickBooks® 2000 its called **Vendor List**).

3. Next, in the **Vendor List** window select the **Vendor** option from the drop-down menu and the **New** option, in the next screen.
4. Next, type the *name* of the company or person in the **Vendor** field at the top of the screen.
5. In the **Company Name** field type the *company name* if the Vendor is a "company". If the vendor is a person skip this field.
6. Next, the following fields may be of secondary importance to you and thus you may elect to skip over them: Mr./Ms./Mrs., First Name, Last Name, the address and the City and the State and the zip code. Further more, the Contact name, Phone and Fax No fields.
7. In the **Additional Info** tab above you may make the vendor default into particular terms such as 2% 10, Net 30, etc. by clicking in the *Terms* field. If this vendor qualifies to receive a 1099 form, you must click in the Vendor eligible for 1099 field. This is one of the requirements in order to be able to print a 1099 form.
8. The **Opening Balance** field allows you to enter the amount you may owe to a new vendor. If you type an amount in this field, the following accounting takes place in the background:

	DR	CR
Uncategorized Expense	$X	
AP account		X

X = Equals the entry amount.

If you <u>do</u> enter an Opening Balance for a new vendor, you must follow through by adjusting the *Uncategorized Expense* account via a journal entry through the *Make Journal Entries* function (please consult with your accountant for the offsetting entry to this adjustment).

Entering an opening balance via the vendor template, as described above, provides only a summary amount. If you prefer detail, you may want to enter an opening balance the way it's described in the *Entering Beginning Balances* section.

9. Next, click the **OK** button to save the information. You have finished creating a new vendor.

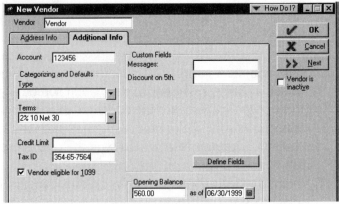

Figure 1.32 shows the information available via the **Additional Info** tab.

Vendor Activity Report

To *view* or *print* a Vendor's **activity** report, select:
1. The **Lists** at the menu bar
2. The **Vendors List** option at the drop down menu
3. *Highlight* the vendor's name
4. Select the **Reports** button below
5. Select the **QuickReport:** option
6. To print, select the **Print** button above

Entering Customers

Customers are companies or people who *buy* your product or services.

To create a *new customer*:
1. Select the **Lists** at the menu bar and the **Customer:Job** option shown below and click once.

Figure 1.33 shows the *Customer:Job* option (in QuickBooks® 2000 it is called **Customer:Job List**).

2. In the Customer:Job window, select the **Customer:Job** from the drop-down menu and the **New** option in the next screen.
3. Next, type the name of the company or person in the **Customer** field at the top of the screen.
4. In the **Company Name** field type the company name if the customer is a "company". If the customer is a person skip this field.
5. Next, fill in the blank fields such as: Mr./Ms./Mrs., First Name, Last Name, Address and etc. The information in these fields can be used later to contact the customers for marketing purposes.
6. Next, continuing by typing the Address, City, State, and the Zip code. Enter next the Contact name, Phone, and Fax numbers. This information can be useful at a later time.
7. **At the Additional Info tab, you may set the following *optional* information:**
   a. The customer *taxable* position by clicking at the **Customer is taxable** field. If you make **it taxable, select the State that you want QuickBooks® to default in.**
   b. Set a *credit limit* if you desire by typing an amount in the **Credit Limit** field. The credit limit is an active field, and if you invoice a customer and the amount exceeds the set credit limit, QuickBooks® will warn you to stop.
   c. Select from the list any *terms* according to your company's policy
   d. Next, you may select a *type* for the customer. The customer types help in the *filtering* of information into reports

8. The **Opening Balance** field allows you to enter the amount the customer owes to your company at the time of creating the customer template. If you type an amount in this field, you will establish an open balance for this customer and the following accounting will take place:

	DR	CR
AR account	$X	
Uncategorized Income		X

X = Equals the entry amount.

If you do enter an opening balance for a customer when you create the template, you must follow through by adjusting the amount that goes into the *Uncategorized Income* account. QuickBooks® uses this account automatically. You can remove the amount from the *Uncategorized Income account* into another account through the *Make Journal Entries* function (please consult with your accountant for the offsetting entry to this adjustment).

Entering an opening balance via the customer template, as described above, it provides only a summary amount. If you want prefer detail, you may choose to enter opening balances the way it is described in the *Entering Beginning Balances* section.

9. Next, click the **OK** button to save the information for the new customer.

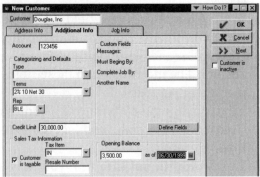

Figure 1.34 shows the information available via the *Additional Info* tab.

Customer Activity Report

The customer activity report known as the QuickReport shows the history of the business you have done with a particular customer.

To *view* or *print* a customer's **activity** report, select:

1. The **Lists** at the menu bar
2. The **Customers:Job** option
3. Next, highlight a customer or a job name
4. Select the **Reports** button at the bottom of the screen
5. Next, select the **QuickReport** option
6. To print the activity report, select the **Print** button at the top of the screen.

Adding Jobs to Customer Names

A job is a project your company is doing for a customer.

After you create the customer in the **Customer:Job List** screen *highlight* the customer name (by clicking once) and select the following options:

1. The **Customer:Job** button
2. In the next screen, select the **Add Job** option
3. Next, type a job *name* at the **Job Name** field. The name can be in *alphanumeric* characters and it identifies the job.
4. Next, click at the **Job Info** tab and type the following *optional* information:
   a. A job status at the **Job Status** field
   b. A job *description*
   c. A select from the list or type a new job *type* at the **Job Type** field as shown in figure 1.35. Please see below for an explanation of the job types.
   d. Select a start date at the Start Date field.
5. Next, click at the **OK** button, and you have finished creating a job for the customer. You can create multiple jobs for a given customer. To create *another* new job for the same customer, repeat the process described above.

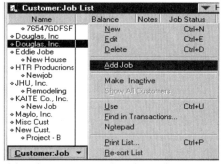

Figure 1.35 shows the new Add Job *option* at the drop down menu.

It is important that you create a Job if you do more than one job for the same customer. Using the Jobs that you create, you can track the *income* and the *expenses* you incur for each job into the *Job Profitability* Report.

Job Types

The job **Types** in QuickBooks® allow you to filter information on the job reports.

For a contractor, Job Types can be for example, the different types of work such as; New Construction, Remodeling, Residential Construction, Commercial Construction, Concrete Walls, Concrete Floors, Commercial Roofing or Residential Roofing, etc..

To create a new **job type**, select the following:
1. The **Lists** at the menu bar
2. The **Other Lists** option from the drop down menu
3. Next, select the **Job Types** option
4. Select the **Job Type** button at the next screen and the **New** option
5. Next, type a *name* for the new job type
6. Click at the **OK** button

To print a job report by *type*, please see the *Job Profitability* report section in Chapter 6.

Entering Beginning Balances

During our previous discussion about the information you will need to start your company, we had talked about the type of information you'll need to start properly in QuickBooks®. In this section, we'll talk about how to enter this information properly.

We have also talked about entering information for customers, vendors and account beginning balances when we described the *Interview* process. Each time we had provided an alternative way of entering beginning information. In this section will discuss another way of entering beginning balances.

The process of entering beginning information the way it will be described in this section, can be *applied* at the very beginning, before you have entered anything in QuickBooks®, or at a later time. For example, it's very possible that you may have been working with QuickBooks® for some time but you have never entered any beginning information. In other words, the system has the information you have been entering since you have started by recording invoices, bills, and payroll but there are no prior balances. These are the balances that existed prior to using QuickBooks®.

It's important that these prior or beginning balances be entered into the system so that when you print the financial statements and other reports, they will be accurate.
Each type of information must be entered through the appropriate function in QuickBooks®.
To describe the process of entering beginning information properly, we'll need to have a specific example. Let's say that after a long process of evaluation, thinking and preparation you have decided to computerize your company and that the *current* date is August 10th.
Based on this example, you have two options to choose from in order to select a *starting* date:
1. You may select January 1st as a starting date or
2. You may select July 1st.

If you select January 1st, there will be more information (more work) to enter but you'll end-up with more *detail* and perhaps more *accurate* information in your system.

If you select July 1st as your starting date, there will be less information to enter but also less detail in your system. Also it maybe a little bit more confusing as you go through the process of managing vendor and customer information.

To make this example simple, let's say you choose January 1st as your starting date. Now, based on this scenario, here is the information you will need:
1. A **Trial Balance** prepared by your accountant as of December 31st as shown in figure 1.36.
2. All the **purchases** you have made from January 1st to the current date separated into:
   a. Paid bills
   b. Unpaid bills
3. All the **sales** your company has made from January 1st to the current date separated into:
   a. Paid invoices
   b. Unpaid invoices
4. All the **checks** written from January 1st to the current date.
5. The **Payroll** total figures accumulated by <u>employee</u> and by <u>quarter</u>.

Here is how to enter the information properly:
1. Enter the **Trial Balance** by recording journal entries through the *Make Journal Entries* function. In chapter 5, we discuss how to make journal entries.
2. Enter the **paid bills** through the **Write Checks** function by recording all the necessary information shush as check number, amount, date, vendor name, etc..
3. Enter the **unpaid bills** through the **Enter Bills** function by recording all the necessary information such as bill number, amount, date, vendor name, etc.. Also, make any necessary memos regarding each transaction.
4. Enter all the **paid** customer invoices through the **Enter Cash Sales** function by recording customer name, check number, invoice number, amount, etc..
5. Enter all the **unpaid** invoices through the **Create Invoices** function. You must create an invoice in QuickBooks® for each manual invoice.
6. Enter **new loans** via the **Make Journal Entries** function the way it is described in chapter 5
7. Enter **loan payments** through the **Write Checks** function just the way we described in the *Manual Check* example, chapter 2.
8. Enter **payroll** information by employee in the following manner:
   A. Paychecks from January 1st to June 30th as described in chapter 1, by selecting:
      a. The **Activities** at the menu bar

b. The **Payroll** from the drop down menu
c. And the **Set Up YTD Amounts** option

B. Paychecks from July 1st to the current date enter one paycheck at a time just the same way we describe the process of doing payroll in Chapter 4. In the Preview Paycheck screen, type the hours and allow QuickBooks® to calculate the payroll for each employee and under the **Amount** column, on the right side, highlight and <u>change</u> the amounts to <u>match</u> the actual amounts of the paycheck you are entering.

To enter the paychecks, select the following:
a. The **Activities** at the menu bar
b. The **Payroll** from the drop down menu
c. And the **Pay Employees** option

International Services, Inc.
Trial Balance
As of July 12, 1999

	Jul 12, '99	
	Debit	Credit
6255 · Postage and Delivery	5.00	
6295 · Rent exp.	1,000.00	
6345 · Telephone	400.00	
Insurance:6410 · Liability Insurance & WC	280.00	
Insurance:6420 · Health Insurance - Group		35.00
6450 · Office Supplies	87.00	
6500 · Commission Exp.	100.00	
6560 · Wages Exp.	570.00	
6570 · Payroll Tax Exp.	71.50	
6999 · Uncategorized Expenses	560.00	
7035 · Other Income		25.00
TOTAL	**307,368.25**	**307,368.25**

Figure 1.36 shows a segment of a **Trial Balance** report.

Note: Please, consult with your accountant about the beginning balance amounts.

Items

𝒫urpose: **Items** help you customize QuickBooks® in a way that will fit your company's needs.

Items add *flexibility* and *power* to QuickBooks® and they are multipurpose.
Here are some of the tasks you'll be able to accomplish through the use of the Items:
- If you are dealing with *product*, you'll be able to record the *purchase* and the *sale* of products.
- If you are a *service* type company, you'll be able to create *estimates* and *bill* customers for the services you provide to them.
- You'll be able to *subtotal*, create *Sales Tax* tables, provide *Discounts*, manage customer *Prepayments*, and etc..

In addition to the tasks mentioned above, Items help you complete the accounting for every transaction they become a part of through the *default* account assigned to an Item.

Items use list
At the table below, we explain the Item *Types* and the *purpose* of using each type:

Table 1

Item Type:	Purpose to Accomplice:
Service	This type allows you to create Items for the services you provide, that you'll use to bill your customers after a service is provided.
Inventory Part	This type allows you to create Items for product you purchase to stock in inventory and later resell to customers. This item will help you record the purchase and the creation of the billing of the customer.
Non-inventory	This type of Item is for product you purchase but do not stock in inventory but you use in customer jobs.
Other Charge	This type allows you to add *other charges* such as freight, finance charge, bad customer check adjustments, penalties, etc. We'll provide specific examples later.

Subtotal	This Item type allows you to calculate a *subtotal* on estimates and invoices.
Group	This Item type allows for grouping together of multiple *Service* and *Inventory Part* type Items for a fast entry of all Items in the group on forms such as Invoices, Estimates, etc.. Individual Items must already exist on the Items list. The Group type does something, remotely, similar to Sub-assembly.
Discount	Calculates an amount as a discount (percent or fixed) and deduct it from the preceding line on an invoice or estimate. (You do not need a **discount item** for discounts on early payments. That type is driven by the Terms)
Payment	This Item type is used to subtract the amount of payment from the total of an invoice received prior to invoicing, so amount owed is reduced.
Sales Tax Item	Calculates sales tax.
Sales Tax Group	This Item type allows for the grouping together of multiple Sales Tax Items. That will allow you to calculate sales tax that is based on one percent amount, and print one check to pay the sales tax but in the Sales Tax Liability report you will have detail by individual Sales Tax Item.

Note: You can create new Items anytime there is a need for a new one.

To create a new **Item**, select the following:
1. The **Lists** at the menu bar.
2. The **Items** option (in QuickBooks® 2000 its called **Item List**).

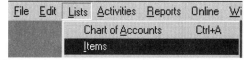

Figure 1.37 shows the *location* of the Items option

3. In the **Items List** screen, click at the **Item** button below and select the **New** option

38

from the drop down list.
4. Next, in the **New Item** screen, select the type of the Item you want to create at the **Type** field. Let's say that the type you want for this example is *Service*.
5. Next, under the **Item Name/Number** field, type a name for the new Item.
6. Under the **Description** field, type a detail description. This description shows up on the description field on estimates and invoices.
7. In the **Rate** field, type an amount i.e. $70, the rate you we'll charge for the service.
8. In the **Taxable** field, you can designate if the Item will be taxable by clicking.
9. In the **Account** field, select an account from the chart-of-accounts. The account selected in this field, will act as the *default* account then whenever you use this Item on an invoice to bill a customer for the service provided, the amount invoiced will end up in the account designated in this field.
10. The field in the middle that's called "**This service is performed by a subcontractor, owner, or partner**" is designed to allow you to make a Service type Item to be used with sub-contractors when you sub a service out. If you click once on it, it opens a new screen with the following fields:
    a. A *Purchase/Sales* description.
    b. A *Cost* rate. That's the rate your subcontractor will charge you for the service.
    c. An *Expense Account*. That's the account that will receive the *cost* of the service when the subcontractor will bill you for services provided.
    d. The *Preferred Vendor* field is designed to allow you to make the Item default to a Vendor's name.
    e. The *Sales Price* field, allows you to enter a rate. That's the *price* you charge for the service.
    f. The *Income Account* field allows you to assign a default *income* type account from the chart-of-accounts so that the income earned from this service (when you invoice the customer) will be recorded into this account.
    g. After you have finished, click the **OK** button to save the new Item information as shown in figure 1.38.

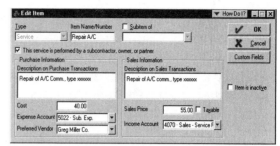

Figure 1.38 shows a completed **Service** type Item.

The same steps can be followed to create other Items of the same or different types. However, we will discuss the creation of an **Inventory Part** type Item next.

To create an **Inventory Part** type Item:
1. Follow steps 1, 2 and 3 from the first example and select the **Inventory Part** type.
2. Type a *Name* in the **Item name/Number** field
3. Type a *Purchase* description and a *Sales* description. The purchase description appears when you use the Item to record a vendor's bill. The sales description appears when you invoice your customers for the sale of product.
4. Type an *amount* at the **Cost** field for the cost of the product and an *amount* at the **Sales Price** field, which is the sale price you'll be charging when you sell the Item.
5. Next, allow the *Account* that appears in the **COGS Account** (Cost of Goods Sold) field to remain. This is a default account and it's the account that will receive the cost of the product at the time of invoicing the customer for the sale of the product.
6. The *Preferred Vendor* field is designed to allow you to make the Item default to a Vendor's name.
7. The *Income Account* field allows you to assign default *Income* type account from the chart-of-accounts so that the income earned from the sale of this product will be recorded into the default account.
8. The **Asset Account** field is to allow you to select an asset type account. This account will receive the incoming inventory when you record a new purchase of product. The proper account is the *Inventory* account.
9. The **Reorder Point** field allows you to enter a quantity of units for this product. QuickBooks® compares the *re-order* field with the *quantity* on-hand. And when the quantity-on-hand goes below the reorder point, QuickBooks® then prints a reminder

in the **Reminders** list to remind you to reorder product so that you do not run out of inventory.

10. At the bottom of the screen, there are three fields: The **Qty on Hand**, **Total Value** and **As of Date**. These fields stay open only for a short time. After you utilize the item one time, those three fields disappear forever. Their purpose is to allow you to establish a quantity on-hand at the time of creating the new Item.

    To establish the quantity on-hand, type in the quantity at the **Qty on Hand** field and simply click at the **Total Value** field and QuickBooks® multiplies the *quantity on-hand* figure with the *cost* figure and establishes the inventory value at the **Total Value** field. The inventory value you establish when you create a new Item results into an actual accounting transaction in the background. QuickBooks® increases the inventory account and the Open Balance Equity account, which acts as a default account. The transaction appears as follows:

	DR	CR
Inventory	$X	
Open Balance Equity		X

    X = Equals the entry amount.

11. After you complete the creation of the item, select the **OK** button to save the new Item information.

    Note: After you have finished creating the new Item, you must follow through and make an adjusting entry where you'll be taking the amount out from the Open Balance Equity account via a debit type entry and a credit entry into another account. Please consult your accountant for the offsetting entry.

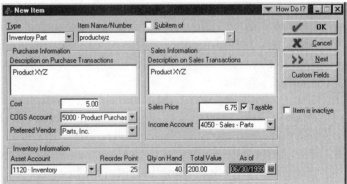

Figure 1.39 shows a new **Inventory Part** type Item.

Another common item is the **Sales Tax** type Item. This Item allows you to track and calculate sales tax. Sales tax is a liability and it accumulates automatically in a liability

41

account that's called **Sales Tax Payable**. You may need to create a separate sales tax Item for your State, County, City, etc. depending on your area's requirements. Please consult with your accountant or you may call your State or local government for more information.

To create a Sales Tax Item:
1. Follow steps 1, 2 and 3 from the first example and select the **sales Tax Item** type
2. Next, type a *name*
3. Next, type a *description*
4. Type a *rate* such as 5 (whatever your states' % rate is) and the percent sign %
5. Select from the vendor list the *Tax Agency* such as the *State Department Of Revenue*. That's the name that will print on the checks when sales tax will be paid
6. Next, click the **OK** button to save the new Item information.

Another common type is the **Subtotal** type Item. This Item type allows you to subtotal figures. It does math for you any time you use it on an Estimate or an Invoice. It creates a subtotal of the numbers positioned above it automatically.
To create a **Subtotal** type item:
1. Follow steps 1, 2 and 3 from the first example and select the **Subtotal** type
2. Next, type a *name*
3. Next, type a *description*
4. Click the **OK** button to save the new Item information.

Another type of an item is the **Other Charge**. This Item type can help you with multiple situations one of which could be to create an Item that would help you record **Customer Prepayments** properly. Customer prepayments are amounts your company may receive from customers for product or services to be provided in the future.

To create an Item to record **Customer Prepayments**, follow these steps:
1. Follow steps 1, 2 and 3 from the first example and select the **Other Charge** *type*.
2. Next, type a *name* that would be fitting of the purpose of this Item such as *Customer Prepayment* or *Customer Deposit*
3. Type a *description* that is fitting to the Item, such as *Less: Deposit Made by your company*.
4. Skip over the *Amount or %* field, and leave the *taxable* field without a checkmark.
5. Select an account to be the default account. This account must be a *current liability*

type that you must create in the chart-of-accounts and name it *Unearned Income*. Note: When recording a customer prepayment, you <u>must</u> avoid affecting an Income type account or the A/R account. Later, under the *Customer Prepayments* section will discuss how to record this type of transaction properly.

6. After you create the item, click **OK** to save the information.

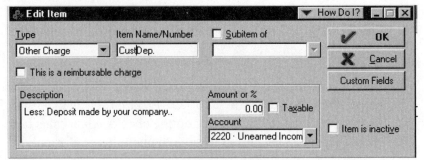

Figure 1.40 shows the **Customer Deposit** (Prepayment) Item.

Another Item type is the **Discount**. By creating an Item of this type, you'll be able to provide <u>discounts</u> to your customers. You can use it when you create Estimates and Invoices by simply clicking on the Item list to select the Item.

To create a new **Discount** type Item, select the following:
1. Follow steps 1, 2, and 3 from the first example and select the **Discount** *type* item.
2. Next, type a *name* in the **Item Name/Number** field.
3. Type a *description*, such as "*Less: Discount of 5%* ".
4. In the **Amount or %**, type a percent such as 5 and the % sign.
5. Then in the **Account** field, select an account that will be the default account. This account can be the *Customer Discounts,* which must be an <u>income</u> type account. Via the use of this Item, you will manage to provide discounts based on *quantity* or *other* purposes to your customers. This item, via the amount that you type in the *Amount or %* field, multiplies the dollar amount of the product or service items by the percent that you specify and allows you to provide a customer with a discount on the invoice or the estimate functions.

Another item that you can find useful is an Item to help you write-off **bad debt**. Bad debt is an amount a customer owes you for services or product you have provided and refuses

to pay you.
Note: Do not be quick to write-off bad debt until you have exhausted all possible avenues of collecting.

To create an Item to manage **Bad Debt**, select the following:
1. Follow steps 1, 2 and 3 from the first example and select the **Service** *type* item:
2. Next, type a *name* that indicates the purpose, such as **Bad Debt**.
3. Type a description such as *Bad debt expense* but do not use the *Rate* or the *Taxable* fields.
4. Next, allow this item to default into the *Bad Debt account* that is expense type.
5. Click at the **OK** button to save the new Item information.

Another use of the **Other Charge** type Item, is to create an Item that would help you correct your data after a customer pays you with a **bad check** (or insufficient funds check).
To manage this kind of situation properly, it requires the use of two items: One item to adjust the checking account, and another to charge back the customer for the bank fee that you have been charged.
A. To create an Item that you can use to correct your data after a **Bad Check**, select the following:
   1. Follow steps 1, 2, and 3 from the previous example and select the **Other Charge** type item.
   2. Next, type a *name* for this Item, such as **Bad Check.**
   3. Type a *description* that is fitting to the Item's purpose and do not use the **Amount** or the **Taxable** fields.
   4. Next, allow the item to default into the **Checking** account (Bank account). By defaulting into the checking account, this item will help you decrease the checking account as shown in figure 1.41.

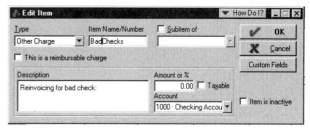

Figure 1.41 shows the **Bad Check** Item.

44

B. The second item that you would need to help you correct your data after the *bad check* situation is also **Other Charge** type. This Item would allow you to *charge back* the customer for the fee the bank has charged your account.
To create this Item, select the following:
1. Follow steps 1, 2, and 3 from the first example and select the **Other Charge** *type* item.
2. Next. type a *name* such as *Bad Check Penalties*
3. Type a *description* such as *Bank fee charge for bad check*.
4. Allow this Item to default into an *Other Income* type account that you could name **Misc. Income**. Through this Item, the bank fee you charge the customer back will be recorded into the Misc. Income account which is reported at the bottom of the P/L statement (because it is not regular income).

Class

*P*urpose: The **Class** (or Classes) function is used as a tracking facility that allows you to track *Income* and *Expenses* into a report that can be used by Management in the decision making process.

The information on the report can be a valuable tool to the manager and helps him/her make better business decisions for the company.

Classes can be used for many different purposes in order to fit your company's needs. Here are some examples:
1. For *Departmentalize* accounting where you track income and expenses by department
2. To track payroll wages by employee *job classification* for the Workman's Compensatio*n* auditor
3. To track the income generated and the expenses incurred by *Salespeople* and etc..

The Class field can be used through such functions as *Write Checks, Enter Bills, Enter Credit Card Charges, Make Journal Entries, Create Invoices, Create Estimates, Enter*

Cash Sales and others. We will be discussing later in more detail as to how the **Class** field can help you track income and expenses.

To create a **new Class**, select the following:
1. The **Lists** at the menu bar.
2. Next, select the **Class** option (in QuickBooks® 2000, select **Lists** and **Class List**).
3. In the Class List window, click once at the **Class** button and select the **New** option
4. In the **New Class** window, type a *name* for a new class
5. Click at the **OK** button.

Terms

*T*erms (or business terms) are the trading conditions that govern business transaction between your company and the customers that buy your product and services, as well as the vendors you purchase their product and services.

To create new **terms**, select the following:
1. The **Lists** at the menu bar.
2. The **Terms** function below (in QuickBooks® 2000, select: **Lists** and **Customer & Vendor Profile Lists**).
3. At the **Terms List** window, select the **Terms** button below and the **New** option.
4. Next, type a **name** for the new terms, such as 2% 10, Net 30.
5. Next, type the number days that payment must be received, such as **30** days at the **Net due in** field.
6. Type the **Discount** percent that would calculate the amount of the discount, such as **2%**.
7. Next, type the number of days allowed for the discount to be applicable, in the **Discount if paid within** field. A commonly used number is **10** days. It means, if payment is received in 10 days, the discount can be taken.
8. Next, click at the **OK** button to save the new terms.

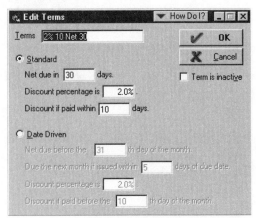

Figure 1.42 shows a 2% 10, Net 30 Terms template.

The **Date Driven** option, is designed to allow you to create terms that are driven by a particular date.

The Accounting Quickhelp Table™

*F*or every business transaction you need to record, you must make sure that you use proper *accounting* and an appropriate QuickBooks® *function*. That's how you can maximize the value of QuickBooks® in your business.

The Accounting Quickhelp Table™ is designed to help you get quick answers in order to achieve both of these objectives mentioned above by listing the most common transactions used by most businesses.

Transactions in Table 2 are listed by transaction type and in an alphabetic order.

Table 2

Type of transaction to record:	QuickBooks® function to use:	Account Name:	Acc. Type:
Bank fee charge	**Register** (for Checking account)	Bank fee charges	Expense
same	**Reconcile** (for checking account)	Bank fee charges	Expense
same	**Make Journal Entries**	Bank fee charges	Expense
Bill from subcontractor (for service)	**Enter Bills**	Subcontractor expense	CGS
Bill from vendor for office supplies	**Enter Bills**	Office supplies	Expense
Bill from vendor for mater. purch.	**Enter Bills**	Material purchases	Expense
C.O.D check (for C.O.D purchase)	**Write Checks**	An expense account depending on purchase	Expense
Check for an employee advance	**Write Checks**	Employee advance	Asset
Check from employee (advance)	**Make Deposits**	Employee advance	Asset
Check to refund for overpayment	**Create Credit Memos/Refunds** next, select the **Refund** button	A/R	Asset
Credit to customer (for product sold)	**Create Credit Memos/Refunds**	Inventory (via the Item)	Asset
		CGS (via the Item)	Expense
		Sales (via the Item)	Income
		A/R	Asset
Credit to customer (for service sold)	**Create Credit Memos/Refunds**	Sales (via the Item)	Income
		A/R	Asset
Credit card credit (from vendor)	**Enter Credit Card Charges**	An expense account depending on the purchase	Expense
Credit card fee charge	**Reconcile** (for a credit card)	Bank/C.C fees	Expense
Credit card purchases	**Enter Credit Card Charges**	An expense account depending on the purchase	Expense
Credit from vendor (prior purchase)	**Enter Bills**	The expense account used for the purchase	Expense
Debit card purchase (debit card)	**Write Checks**	An expense account depending on the purchase	Expense

Electronic fund transfer (electronic purchase from checking account)	**Write Checks**	An expense account depending on the purchase	Expense
Employee purchase on co. account (record it when the vendor sends you the bill)	**Enter Bills**	Employee advance	Asset
Employee reimbursement check	**Write Checks**	An expense account depending on the reimbur.	Expense
Invoice customers	**Create Invoices**	Sales (via the Item)	Income
Loan payment check (bank loan)	**Write Checks**	Interest expense (interest)	Expense
		Loan account (principal)	Liability
Pay credit card	**Write Checks** (after reconciling)	Credit card account	Liability
	Pay Bills (after reconciling)	Credit card account	Liability
Pay employees	**Payroll**	Wages/Salaries expense	Expense
Pay the sales tax	**Pay Sales Tax**	Sales tax payable done by QuickBooks®	Liability
Pay outstanding vendor bills	**Pay Bills**	A/P and Checking account done by QuickBooks®	
Payments from customers	**Receive Payments**	A/R and Checking account done by QuickBooks®	
Prepayment from customer (for work to be done in the future)	**Enter Cash sales**	Customer deposits	Liability
Purchase discounts (paying early)	**Pay Bills**	Purchase discounts	Income
Refund check- From Insurance	**Make Deposits**	Insurance liability	Expense
Refund check- from Payroll taxes	**Make Deposits**	A liability account that the tax was paid from	Liability

Chapter 2

Purchases & Payments

$\mathcal{T}$he purpose of this chapter is to help you learn how to record the various types of *purchases* your company makes and the *payments* to Vendors. We'll discuss the purchases made using cash, credit cards and on account.
We'll talk about how to work properly with credit cards, how to manage trade debt and about the business practices that pertain to this subject.

In this chapter we'll examine the following QuickBooks® functions:

- **Write Checks**
- **Petty Cash**
- **Transfer Funds**
- **Create Purchase Orders**
- **Enter Vendor Bills**
- **Credit Card Charges & Credits**
- **Pay Vendor Bills**
- **The Register**
- **Pay Sales Tax**

Write Checks

*P*urpose: To create **checks** for disbursement type business transactions.

A disbursement is a business transaction through which you are able to do two things in one step: 1. Record the purchase of a product or service and
2. The payment (by printing a check).

Disbursements are recorded properly through the **Write Checks** function. Through this function you can record a number of different transactions that pertain to your business.

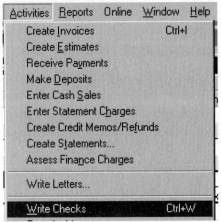

Figure 2.1 shows the **Write Checks** function (in QuickBooks® 2000, select **Banking** and **Write Checks**).

In the examples that follow, we'll be discussing specific **disbursement** type transactions that you can apply to your own company.

Example #1: printing a check for a purchase that is based on **C.O.D** (charge on destination).
The C.O.D, is a prime example of a disbursement type transaction. It is a transaction that demands payment upon delivery. For example, let's say that your company has ordered materials (or product) from a vendor with whom you have no prior business history and the only way to make this purchase is if you agree to pay C.O.D. When the merchandise arrives, you must hand over a check and the delivery person gives you the material you have ordered.

To record this transaction (the purchase and payment) and print the check, select the following:
1. The **Activities** at the menu bar
2. The **Write Checks** option (in QuickBooks® 2000, select **Banking** and **Write Checks**).
3. At the **Pay to The Order** of field select the name of the vendor from the vendor list.
4. Next, either press the **Enter** key or click with the left **mouse** button to move to the next field.
   If a particular name is not in the system, QuickBooks® will prompt you to enter it.
   To enter the name, you have two options: **Set Up** and the **Quick Add** option:
   a. Via the **Set Up** option, you may enter all the necessary information such as the address, the City, State, zip, telephone number, fax number, a contact, etc..
   b. The **Quick Add** option retains only the name of the vendor (company).
   To proceed, click at the **Quick Add** button. Next, select the *type*, which in this case is vendor. Next, click at the **OK** button. Now the new name is in the Vendor name list.
5. After the name is added, instantly the cursor moves to the **Amount** field. That's where you type the *amount* of the purchase. The amount can be typed on the check (upper portion of the screen) or at the lower part of the screen where you have two tabs:
   a. **Items** and
   b. **Expenses**.
   The **Items** option is designed to allow you direct access to the Items listing for the purpose of recording the purchase of *product* or *materials* that you don't intend to utilize immediately. In other words, if you buy product or material that you'll be storing and <u>not</u> use immediately, you need to use the **Items** tab so the you can record the purchased product into the **Inventory** account, which is an asset type account. In QuickBooks®, you can manage inventory via the use of an Item that is type **Inventory Part** (as explained in Chapter 1, in the Items section). Also, via the Items tab, you may record the purchase of a service that a Vendor may bill you.

   The **Expense** option allows you to <u>expense</u> the purchase of material (or product) that you have purchased and intend to utilize fairly soon. **Any time you buy something with the intent to utilize immediately you must expense it**. Expensing is accomplished through the **Expense** option and via an expense type account from the *chart-of-accounts*. Also, via the Expense tab, you may record the purchase of a service that a Vendor may bill you.
   When you click under the Expense column, it opens the chart-of-accounts, which

automatically defaults to the expense account section. Select the appropriate account you want to use by clicking.

6. Next, type the amount in the **Amount** field. Let's say the amount of this purchase is $500.
   Note: If you type the amount in the check section and you press the Enter key or click anywhere, the amount will instantly go also in the lower section of the screen (under the Expense or Items options). The same would have happened if the amount were first typed in the lower section. It would have gone up in the check portion. Therefore, you may only type the amount once.

7. Next, you may optionally use the **Memo** field. Any text you type in this field will print in the expense account's *Quick Report* and the *General Ledger*. The **Memo** field at the bottom of the check however, will print on the *check*, in the *Register* of the Bank account and in the vendor's *Activity* reports.

Now, going through all these steps, you have managed to record this transaction and complete the following accounting: The amount in the check represents a *credit* type entry, and that means it will *decrease* the checking account (where monies come out from). The amount under the Expense field, is a *debit* type entry, and that *increases* the Expense account selected next to it (actually the amount has been split into $300 and $200, as shown in figure 2.2, to accommodate job costing and departmentalized accounting, which we'll explain later).

Here is the actual accounting that took place in the background:

	DR	CR
Material purchase account	$300	
Material purchase account	$200	
Checking account		500

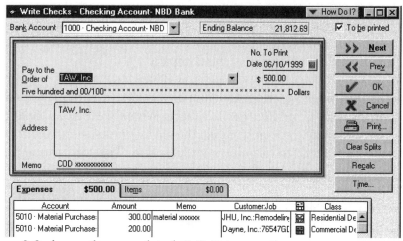

Figure 2.2 shows the completed **C.O.D** transaction.

Therefore, the <u>purchase</u> has been recorded and the <u>check</u> has been created. You've accomplished two things in one step.

8. Next, you may proceed by selecting the following:
    a. Click at the **To be printed** field to print later (via the print batch) and the **OK** button to post and close the function.
    b. Click at the **To be printed** field to print later (via the print batch) and the **Next** button to post and proceed to create another check.
    c. Click at the **Print** button (to the right) to print the check immediately.

However, in order to use QuickBooks® to its full potential, do not make a selection in step # 8 because you need to use the rest of the functions available to you so that you can maximize the value of QuickBooks® to your company.

QuickBooks® is an excellent business tool. The more you use it, the more you can expect to get out of it in the form of printed reports. So, with that in mind, we will talk about the **Customer:Job** and the **Class** fields as shown in figure 2.2.

Job Costing Expenses

The **Customer:Job** field is intended to allow you to do what is known in the business world as *Job* (or Project) *Costing*. It helps you to track *income* that is earned and *expenses*

54

your company incurs from a Job into the Job report. This report is a *Profit and Loss* or *P&L* report and its purpose is to show the manager how the company is performing, whether it is making or losing money, on a per Job basis. Therefore, Job Costing is very important to the success of the company.

You may track *expenses* the company incurs from the *Write Checks*, *Enter Bills* and *Enter Credit Card Charges* functions, which we'll examine at a later point in this book. Also, you may track *income* via the *Create Invoice* and the *Enter Cash Sales* functions again, we intend to examine a bit later also. The Job costing report is a report you should print on a weekly basis.

To do **Job Costing**:
1. Click on the arrow button under the **Customer:Job** column, as shown in figure 2.3, and select a customer or job name from the Customer:Job list.
   *Note: To create a new job for a customers, see the **Entering Customers** section, in Chapter 1.*

Expenses $500.00	Items	$0.00		Time...
Account	Amount	Memo	Customer:Job	Class
5010 · Material Purchase:	300.00	material xxxxxx	JHU, Inc.:Remodelin;	Residential De
5010 · Material Purchase:	200.00		Dayne, Inc.:76547GI	Commercial De

Figure 2.3 shows the Job *names* selected from the Customer:Job list.

You first must create the customer in the customer list and if your company has more than one job going on, at the same time, for the same customer, you need to create a job name for each job, under the customer name. The Job then becomes an entity through which you may *track* expenses and income, into the job report, which is called the **Job Profitability** report.

2. Next, click at the **OK** or the **Next** button to post the expense into the job report.

   To print the **Job Profitability** report, select the following:
   a. The **Reports** at the menu bar
   b. Next, the **Project Reports** (in QuickBooks® 2000, select **Reports** and **Jobs & Time**).
   c. Next, select the **Job Profitability** report.
      This report comes in a *summary* or *detail* format. Profitability simply means that it's a *profit* and *loss* report and it provides you with very useful business

information: It shows you whether you company is making (or loosing) money on a per job basis.

In the event that the purchased material were *split* between two or more projects, you can do multiple job costing by splitting the amount of the purchase as shown in figure 2.3 above.

For example, from the total amount of the $500 purchase, you can track $300 in one project and the balance of $200 you can track into another project. All you have to do is under the **Account** column select (by clicking) another account from the chart-of-account list, and under the **Customer:Job** column, select another job *name* from the list in order to track the balance of the purchase amount.

Billable Expenses

*P*urpose: To make an expense such as material, supplies or subcontractor expenses as a **Billable** expense that you eventually will bill your customer.

A billable expense is a purchase you have made for a particular Job (or project) you are currently working on and that you have an agreement that allows you to bill the customer for *material* you have purchased and perhaps the *time* spent working on the Job. This type of an agreement is called **T&M** (time & material).

Please notice in figure 2.4 that next to the Customer:Job field, there is an **Icon**. This Icon plays a key role in QuickBooks®. It allows you to designate an expense as *billable*.

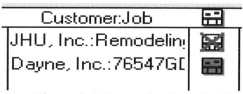

Figure 2.4 shows the **Icon** field.

For example, let's say your company is currently working on a project for customer Dayne, Inc., as shown in figure 2.3, with an agreement that is based on T&M, and that out of the $500 purchase, $200 worth of the material was utilized in the Dayne project. That means you must designate the $200 to be a *billable* expense (you'll bill the customer back for this expense).

To make an expense billable to a customer, all you have to do is leave the icon intact. Just the way it has defaulted. Figure 2.4 above shows the Dayne project as billable. The instant you proceed with the posting of this transaction, the purchase amount designated as billable, will move into the billable list, which is in the **Create Invoice** function (we'll discuss it next).

Figure 2.4 shows the *billable* Icon for the Dayne project. It also shows the JHU project with an X on the Icon, which means that this particular expense is not billable.

After you designate an expense as billable, there is a **report** available that allows you to monitor *unbilled* transactions.

To print the *unbilled expenses* report, select the following:
1. The **Reports** at the menu bar
2. The **A/R Reports** (in QuickBooks® 2000, select **Reports** and **Customers & Receivables**).
3. The **Unbilled Costs by Job** at the drop down menu

To **bill** a customer for billable expenses, select:
1. The **Activities** at the menu bar.
2. The **Create Invoices** option (in QuickBooks® 2000, select **Customers** and **Create Invoices**).
3. Select a customer name at the **Customer:Job** field.
4. Next, click at the **Time/Costs** button to the right of the invoice.
5. In the next screen, select the **Expense** (or the **Items**) tab
6. Next, click once under the **Use** column to place a checkmark (√)
7. Click the **OK** button to create the invoice.

On the other hand, if your company is working on a project that is based on a *fixed* price due to a contract, you can't bill the customer additionally for the expenses you incur.

To make an amount as non-billable, simply point on the **Icon** and click once on it. That will place an **X** on the icon and will remove the amount from being a billable expense as shown in figures 2.3 and 2.4.

Using Classes

*P*urpose: The **Class** field is a tracking facility that can be used to track *income* and *expenses* into a report.

Classes are located next to the Icon field as shown in figures 2.3 and 2.5. It is another valuable business tool that QuickBooks® provides and it can be used for various purposes depending on a company's needs.

Figure 2.5 shows the position of the **Class** (and Customer:Job) field.

One possible use of the Class field is to do **departmentalized** accounting. If your company has multiple office locations, you may define each of these locations as a *department*. A department is simply a section of your business. Also, multiple lines of businesses that may exist within a company can each be identified as a *department*. For example, in figure 2.5 it shows that this company operates in two sectors of the economy: In the **Commercial** and the **Residential** sectors. So, in this case there can be a department (or class) created for each sector.

Note: To create a new **Class**, please see the section about **Class** in Chapter 1.

After creating the departments, you can track *income* the company generates and *expenses* it incurs into a P&L type report. This report will show the manager the company's operating performance on a per *department* level.

To print the **P/L by Class,** select the following:
1. The **Reports** at the menu bar (in QuickBooks® 2000, select **Reports** and **Company & Financial**).
2. Next, select the **Profit and Loss** option at the drop down menu.
3. The **By Class** option (in QuickBooks® 2000, select **Profit & Loss by Class**).

The **P&L** by Department (or Class) report provides valuable information because it divides the company into smaller sections and thus, it shows greater detail. It allows the manager to make better decisions.

To track expenses into **Classes**, select the following:
1. Click on the arrow button under the **Class** column and select the Class you desire to track next to each respective amount. In our example, $300 is tracked into the residential department and the $200 into the commercial department as shown in figure 2.3.
2. Next, click at the **OK** or the **Next** button to *post* the transaction and track the amounts into the Class report.

Another possible purpose of the **Classes** is to help you track the company's payroll into *classifications* for the Workman's Compensation auditor. For example, you may create Classes that identify your company's employee positions into various classifications such as Office, Field, Shop and etc. Once you create the Classes via the *Payroll* you may track the company's wage classifications into the report.

Another possible use of the **Class** field is to create a Class for each one of your *sales* people simply by using the salesperson's name. By using the Class you have created, you may track the *expenses* they incur and the *income* they generate for you, and then you can print a P&L report by sales person. This report will show you the profitability of each sales person.
Based on your knowledge of QuickBooks® and your imagination, you can come up with all kinds of new ideas as to how this function can be used to track valuable, additional information, for your particular company's needs

Printing Checks

To print a check in QuickBooks®, there are two options available to the user:
A. Batch printing:
1. To print via the <u>batch</u> printing click at the **To Be Printed** field on the upper right corner as shown in figure 2.6. Next, proceed by clicking at the **OK** or the **Next** buttons and instantly QuickBooks® will *post* the transaction and *create* the check.

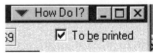

Figure 2.6 shows the "**To be printed**" field (in QuickBooks® 2000, this field is under the check).

The **To Be Printed** field, sends checks (and other forms such as invoices, paychecks, credit memos, etc.) you create into the *batch* printing. The batch printing is an intermediate step and it is intended to allow you to print multiple forms at once instead of one form at a time. Of course if you need to, you may print one form at a time via the batch printing too.

2. Next, select **File** at the menu bar
3. The **Print Forms** at the drop down menu
4. The **Print Checks** option as shown in figure 2.7 (in QuickBooks® 2000, its **Checks**).

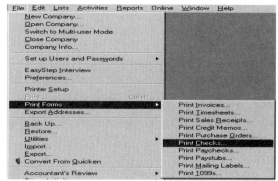

Figure 2.7 shows the location of the *print* batch option.

5. Select the check (s) you want to print by clicking once to place a (√) next to it. Just like figure 2.8 shows.
6. Click at the **OK** button to begin printing.

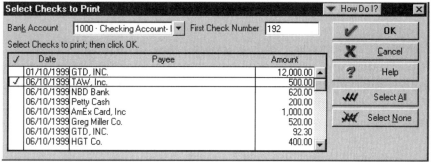

Figure 2.8 shows the *print batch* screen.

B. Print button:

The **Print** button, as shown in figure 2.9 at the upper right corner, provides another option for printing. When you click on it, it sends the form (Check) directly to the *printer*

60

driver instead of the *batch printing*. It is intended to allow you to print one form at a time. To print via the Print button, proceed by:
1. Clicking at the **Print** button
2. Next, click the **OK** button
3. At the next screen click on the **Print** button to complete the printing.

Figure 2.9 shows the **Print** button location.

Before you proceed with printing from either function, you need to make sure you will be using the right *check number*. The purpose of that is to insure that the amounts will match with the check numbers that appear on the preprinted form so that when you go through the process of reconciling the bank statement later everything will be in order. Check *numbers* and *amounts* will match.

Example #2: Recording an **Electronic Fund Transfer** (EFT).
If you authorize a vendor to withdraw monies directly from your account at the Bank for the payment of product or services, you can record these types of transactions through the **Write Checks** function. Of course, when monies are deducted from your account at the bank you don't have to *print* a check.

To record an **Electronic Fund Transfer** transaction:
1. Follow steps 1 & 2 from example # 1
2. Next, select the vendor name at the **Pay to the Order of** field.
3. Select the proper *date* at the **Date** field.
4. Next, remove the checkmark (√) from the **To Be Printed** button.
5. Next, highlight the **Number** field and type a <u>custom</u> set of alphanumeric characters

such as EFTIP7999 (this custom number indicates: Electronic Fund Transfer, the month of the transaction, the date and the year plus any other characters you may want to type).

6. Next, select the **Expenses** option and click under the **Account** column for the chart-of-accounts. Select the *insurance expense* account.
7. Next, type the *amount* in the **Amount** field. For example, let's say the amount in this case is $140 (which we'll split into two amounts for the purpose of doing *Departmentalized* accounting as shown in figure 2.10).

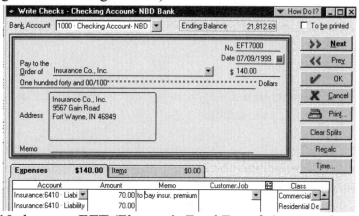

Figure 2.10 shows an **EFT** (Electronic Fund Transfer) transaction.

Remember: Once you begin with departmentalized accounting, you have to be consistent by tracking income and expenses so that the P/L by Class (department) will be accurate.

8. Next, in the **Class** select the appropriate *department* (Class) from the Class list to track the expense by department. For this particular example job costing is not applicable, so we skip the Customer:Job field.
9. Click at the **Next** or **OK** to post and complete recording this transaction.

The accounting that has taken place upon completing the recording of this transaction, is as follows:

	DR	**CR**
Insurance expense account	$70	
Insurance expense account	$70	
Checking account		140

62

If you have a **Debit Card** from your bank, you can record purchases you make using the Debit Card in the same way as in the EFT example above, except that in step # 5 you may want to type a set of characters that identify the Debit Card transactions. For example, let's say that you've purchased supplies with your Debit Card on August 20, 1999. To record this transaction, you may want to type the following at the **Number** field on the check: DC82099 (this custom number identifies the debit card as DC, the month, the date and year when the transaction was done).

Memorizing Transactions

𝒫urpose: To place in the system's memory transactions that will be repeating in the future in order to reduce the amount of work that will be needed to record the same transactions in the future.

Transactions that can be memorized are Checks, Customer Invoices, Vendor Bills, Journal entries, etc. The idea again is to reduce the amount or time needed to complete the same task in the future. The **Memorize Check** function can help you increase your efficiency in using QuickBooks®.

To **memorize** a disbursement check such as the example above, after you complete the information fields on the check, select the following:
1. The **Edit** at the menu bar
2. The **Memorize Check** option as shown in figure 2.11, by clicking once on it

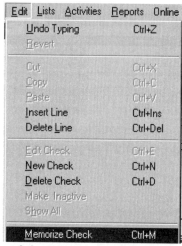

Figure 2.11 shows the location of the **Memorize Check** function.

3. Next, type a *name* in the **Name** field. The Memorize function picks up the vendor name automatically. However, you may type another name if you wish.
4. Next, select the type of memorizing you want such as **Remind Me** or **Automatically Enter**:
   a. The **Remind Me** option repeats the transaction on a semi-automatic basis and it uses the *Reminder*s list. Because it's semi-automatic it allows you to *edit* a transaction. Editing in the computerized environment language means *changing*.
   b. The **Automatically Enter** option works automatically and allows no editing.
5. At the **How Often** field, select the *frequency* by clicking. The frequencies are pre-set in QuickBooks®.
6. Next, select a date at the **Next Date** field. That will indicate the date you want this memorized transaction to begin posting i.e. a month later.
7. Select the **OK** button to make the check a memorized transaction.
   You may follow the same process to memorize vendor Bills, customer Invoices, Credit Memos, journal entries (Make Journal Entry) and etc.

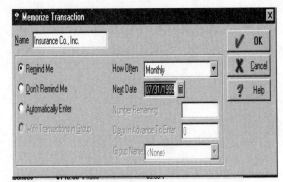

Figure 2.12 shows the **Memorized Transaction** screen.

Once you memorize transactions, you automatically create a list of the Memorized transactions.

To *view* or *print* the list, select the following:
a. The **Lists** at the menu bar.
b. The **Memorized Transactions** (in QuickBooks® 2000, its **Memorized Transaction List**).
c. Select next the **Memorized Transaction** button (at the bottom).
d. In the next screen, click once at the **Print List** option. That will print a listing of the memorized transactions.

The **Remind Me** type option enters the memorized transaction into the **Reminders** window that comes up when you start QuickBooks® (please note: you must activate the **Reminder** preference first in **Preferences**), which is listed under the heading of *Memorized Transaction Due* as shown in highlight mode in figure 2.13 below.

Due Date	Description	Amount
	Money to Deposit	271.88
	Bills to Pay	-860.00
	Overdue Invoices	16,404.93
	Checks to Print	-18,944.87
	Paychecks to Print	-246.95
	Invoices/Credit Memos to Print	13,859.24
	Sales Receipts to Print	271.88
	Purchase Orders to Print	-3,180.00
	Memorized Transactions Due	**-140.00**

Figure 2.13 shows the **Reminders** list screen.

65

To activate a previously Memorized transaction from the **Reminders** window:
1. Click twice on the **Memorized Transaction Due** line in the **Reminders** window. That will open all the transactions that have been memorized.
2. Select the transaction you want to activate by clicking *twice* on it.
3. Next, click once on the **Enter Transaction** button in the next window. The original transaction opens up for you. In case you need to change an amount, you may highlight and change it. If you don't need to change (or edit) an amount, simply click the **OK** button and that will *post* the transaction.
Once you post a Memorized transaction, the **date** moves up to the next month. So that it will remind you again, in the same manner, via the **Reminders** window, in the next month.

The **Automatically Enter** memorized type works automatically. When you open QuickBooks® the start up window automatically pops up to remind you of the transaction that needs to be entered. At the start up window, you will have two options (two buttons): **Now** and **Later**.
1. Clicking at the **Now** button it instantly posts the transaction and creates a check to be printed if it was set to do so (or an Invoice, record a Bill, etc.).
2. Clicking at the **Later** button, will stop the process and will bring up the Start Up window the next time you start up QuickBooks® again.

Example #3: Creating a *check* for an installment payment of a **Bank loan**.
Creating a check for an installment payment every month, is a disbursement type transaction and it must be done via the **Write Checks** function.

It is important to remember that for every loan your company may have with a bank (s), you must create an account for it in the chart-of-accounts. If a loan is a short-term loan, the account you create must be **Other Current Liability** type. On the other hand, if the loan is a long-term the account must be **Long-Term Liability** type.
Business tip: A long-term loan is a loan with a maturity date of twelve or more months.

For example, let's say your company borrowed $30,000, at 9% for 5 years from the Bank to purchase a truck. Based on the scenarios above, monthly payment is $622.70 (including interest and principal) with the following split: $221.92 for interest and $400.78 is for principal.

To create a check for the **payment** of a loan:
1. Follow the steps from example # 1.
2. Select the *name* of the Bank, from the vendor list, at the **Pay to the Order of** field.
3. Type the *amount* of the loan in the **Amoun**t field on the check. In this case it's $622.70.
4. Select the *date* to be the date of creating the check.
5. Click in the **To Be Printed** field on the upper right corner, indicating that this check is to be printed.
6. At the **Expenses** field below, from the Chart-of-Accounts, select two accounts:
   a. Select the *Interest Expense* account and type next to it the **interest** portion of the payment. In this case $221.92.
   b. Next, select the *liability* account of the loan (that's the account where the loan was recorded originally) and type next to it the amount of the **principal**. In this case it's $400.78.

   The accounting from this example is the following: the $622.70 on the check is a *credit* type entry, and it will reduce the checking account. The $221.92 is a *debit* entry and it will increase the interest expense account. The $400.78 is also a *debit* entry and it will decrease the liability account.

	DR	CR
Interest Expense	$221.92	
L-T loan - Bank X (liability)	$400.78	
Checking Account		622.70

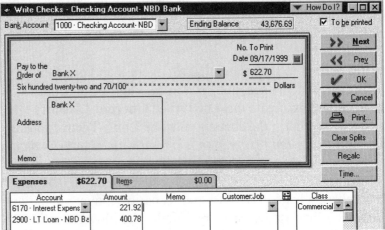

Figure 2.14 shows a disbursement **check** for an installment payment for the bank loan.

7. Next, avoid using the **Customer:Job** field because this transaction does not apply to a job.
8. At the **Class** field select a class from the list to track the cost of the *interest* into a *department* (or Class) to complete the departmentalized accounting.
   Business tip: Do not track Asset, Liability and Equity type entries for Job costing and Departmentalized accounting purposes as we have discussed above. You may track only <u>Income</u> and <u>Expense</u> type entries.
9. Next, click the **OK** button to post the transaction and create the check.

Loan payment checks can be *memorized* but remember to select the **Remind Me** option in the Memorize Check function so that you'll be able to edit the amount of the interest and the principal because they will vary every month. Memorizing checks that will repeat in the future will save you a lot of time because you don't have to go through all the steps again.

Example #4: Creating a check to **Reimburse** an employee for mileage or other misc. reimbursements.
Business tip: In order to follow proper business practices and make transactions easier to record, do not advance employees with company cash. Encourage employees to use their own resources, such as their cash, checks, or their credit cards to buy material or supplies for the company. When the employees submit their receipts you can print checks to reimburse them.

Employee reimbursement checks can be created through the **Write Checks** function. Also, you can reimburse employees through the **Payroll** function when you create paychecks to pay their wages or salaries.

To continue with creating a check via the **Write Checks** function, let's say that an employee provides you two receipts and asks that you reimburse him/her. The receipts are as follows:
   a. One is for *gas* that has been used for the company truck.
   b. The other receipt is for *supplies* purchased for one of the projects your company has currently been working.

To **reimburse** an employee:
1. Follow step 1 & 2 from example # 1
2. In the **Pay to the Order** field select the employee's name from the Employee name list. QuickBooks® will warn you whether you may want to create a paycheck for this

employee. Click at the **OK** button to proceed because you do not want to create a paycheck.
3. Next, make sure that the **To Be Printed** field has a checkmark (✓) because you need to print the check.
4. Select the proper *date* for the check.
5. Type the *amount* in the **Amount** field on the check.
   _The QuickBooks® **Calculator**:
   In this example, the employee handed you two receipts: $20.00 for gas and $70 for supplies. When you have to work with multiple amounts you can activate the built-in QuickBooks® **calculator** in order to add the amounts. To use the calculator, type the first amount in the Amount field and next, press the **Plus** function key. That instantly activates the calculator. Next, type another amount and so on (amount and +. Amount and +. That's the sequence). When you have finished typing the last number, press the **Enter** key to sum all the numbers. You can use the QuickBooks® calculator for as many different amounts as you want.
   The calculator works with the function keyboard keys (+, -, *, /) and can be used for *adding*, *subtracting*, *multiplying*, and *dividing*. For example, to add, click the *Plus* function key. To subtract, hit the *hyphen key*, to multiply use the *asterisk*. To divide, use the forward *slash*.
6. Next, select the **Expenses** option to expense this purchase:
   a. Select the expense account **Auto Expense account** to record the expense of $20 for gas.
   b. Select the **Supplies account** to record the supplies expense of $70.
7. Next, click at the **Customer:Job** field and from the list select the appropriate Job name in order to track the expense of the supplies into the Job report.
8. Next, click on the **Icon** once (x-it) to remove it from the billable list in the event that this expense isn't billable.
9. Next, click under the **Class** column to do departmentalized accounting and track this expense of $70 into the proper Department.
   And that completed the recording of an employee reimbursement.
10. Next, click at the **OK** or the **Next** button to post the transaction.
    The accounting completed for this transaction is as follows:

	DR	CR
Auto Expense	$20	
Misc. Supplies	70	
Checking Account		90

Example #5: Recording a **Manual Check.**

Business people often go to a vendor to purchase product or material, having a company check with them. They buy the supplies or materials and pay with the company check at the counter.

Soon after this type of purchase is made, the bookkeeper must receive all the information in order that the transaction is recorded properly in the system. That's called recording a **Manual Check.**

To record a **manual check** follow all the steps described in the prior examples, except:
1. At the **To be Printed** field, remove the checkmark (✓) to deselect the printing.
2. In the **No** field, type the *actual* check number written at the time of the purchase (the check that has already been issued manually).
   For example, let's say the boss has purchased material that had been stored in the warehouse. Because the material was not used, you must record this purchase of product into the **Inventory** account, which is an *asset* account as shown in figure 2.15.
   To accomplish this task, you need to select the **Items** option, at the lower section of the screen, and proceed by selecting:
1. The appropriate **Item** from the Item list. The Item you need to record the addition of product (or material) to the inventory is **Inventory Part** type. Select the item by pointing and clicking.
2. Next, under the **Qty** field, type the *quantity*. Let's say it's 10. When the Item was created in QuickBooks® at a prior time the cost was set at $6.50. But now lets say, the boss paid only $3 each. So, type $3 in the **Cost** field and instantly QuickBooks® calculates the purchase amount which is $30 ($3 X 10).
3. Next, click at the **OK** or the **Next** button to post the transaction.

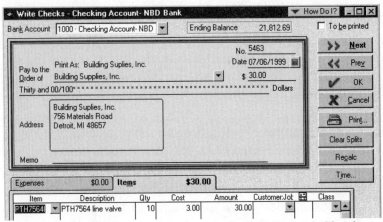
Figure 2.15 shows the recording of a purchase using a **Manual** Check.

As soon as this transaction is posted, the following accounts will be affected:
a. The **checking** account will decrease by $30
b. The value of **inventory** will increase by $30.
c. The quantity on hand of the item purchased will increase by 10.

And here is the accounting that took place in the background from this transaction:

	DR	CR
Inventory account	$30	
Checking account		30

The process described in the above example of adding material or product to inventory is how you can add purchases of product or material to the **Inventory** account. QuickBooks® manages inventory by using the *perpetual* method of inventory management. That means, when you buy product and record the purchase in the system, you'll increase the $ *value* and the *quantity* on hand of the product. When you create an Invoice to bill a customer for the sale of product, QuickBooks® decreases the $ *value* of the *inventory* asset account and the *quantity* on hand. Furthermore, QuickBooks® uses the *average* method of inventory evaluation to figure out the value of your Inventory. The value of inventory appears on the asset section of the Balance Sheet.

Use of Inventory

In the business world, often there is the need to carry in inventory large quantities of product (or material) that will eventually be used towards the completion of various Jobs or projects. These inventoried materials or product cannot be billed separately to the customer on an invoice because they are part of the cost that was estimated to complete the project. Because you can't use the Create Invoices function through which the inventory account can be adjusted (when you use an *Inventory Part* type Item in the *Create Invoice* function, QuickBooks® automatically decreases the inventory asset account), you need to make a manual adjustment the way we'll show you in this section.

To **adjust** the inventory for materials used in a Job, follow the steps outlined below:
1. Select the **Activities** at the menu bar (in QuickBooks® 2000, select **Vendors**).
2. Next, select the **Inventory** option (in QuickBooks® 2000, select **Inventory Activities**).
3. Select the **Adjust QTY/Value on Hand** option, as shown in figure 2.16.

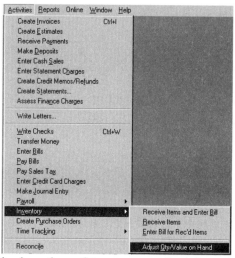

Figure 2.16 shows the location of *inventory adjust* function.

4. Select the proper *date* at the **Adjustment Date** field
5. Type a *reference #* at the **Ref. No** field.
6. Select an expense account such as the *Material Expense* by clicking at the button in the **Adjustment Account** field.
7. Select a job name at the **Customer:Job** field to track the cost of the inventoried material to a particular project. Selecting a **Class** for departmentalized accounting is

optional.
8. Next, click at the **Value Adjustment** field so the value will be calculated.
9. Under the **New Qty** field, click next to the item and type the <u>new</u> quantity that is left in inventory. For example, if quantity was 400 and you have used 100, type 300, that's the new quantity.
10. Click at the **OK** button to post the transaction.
    This process described above will decrease the inventory *value* and the quantity on hand, and it will *increase* the expense account selected to complete the transactions.

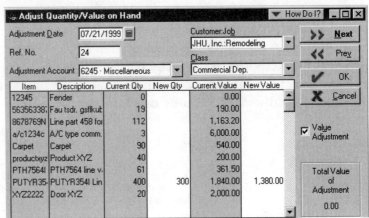

Figure 2.17 shows the *adjustment* of inventory.

The accounting that is accomplished through this example, is as follows:

	DR	CR
Material expense	$100	
Inventory account		100

Editing, Voiding and Deleting

In the event that you made a mistake while creating a check (or another transaction), you can correct it (or edit it), by clicking the **Previous** or **Next** buttons to the right of the *check* screen in the **Write Checks** function until you find it. After you find it you may proceed to correct it by highlighting and typing over the new amount.
Another way to correct transactions is through the *chart-of-accounts* and next, the

73

account **Register**.
To use the **Register** for editing a transaction, select the following:
1. The **Lists** at the menu bar and
2. The **Chart-of-Accounts** and click once on it.
3. Select the *account* where the transaction has been recorded and you wish to change (or edit), highlight it and click twice on it. Or, highlight and click the **Activities** button below. Next, select the **Use Register** button and click once to open the account's **Register**. At the account **Register**, you may click once at the **1-Line** button at the lower right section and that will allow you to view the Register in a *one* or *two* line format.

Figure 2.18 shows how to **open** the Register via the *Use Register* button.

4. Locate the transaction in the Register, highlight it and click the **Edit** button below (in QuickBooks® 2000, its **Edit Transaction**), as shown in figure 2.19, to open the transaction. In the actual transaction, *highlight* the amount you want to change and *type* the new amount.
5. Next, proceed to click on the **Recalc** button ((in QuickBooks® 2000, its **Recalculate**) on the right side, to recalculate the transaction and assure that you are in balance.
6. Click on the **OK** button to re-post the transaction. Now the transaction has been changed in the Register as well.

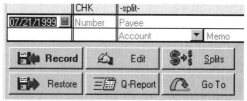

Figure 2.19 shows the lower section of the *Register*.

In the event that you have a transaction that you don't wish to keep, you may remove it by using the **Void** or the **Delete** options. Voiding is an acceptable way to remove a transaction. Deleting is *not* acceptable because it removes every trace of the transaction from the Register and it doesn't allow an audit trail.

To **void** a transaction:
1. *Highlight* the transaction and
2. Select **Edit** at the menu bar, and the **Void** option below. The minute you click, instantly the transaction zeros out, meaning all accounts affected go back to their prior balance. Also, it adds the word *Void* in the memo field. After you have voided the transaction, select by clicking at the **Record** button to record the voiding of the transaction.

The **delete** function removes the transaction from the system as described previously.
To **delete** a transaction:
1. *Highlight* the transaction and
2. Select **Edit** at the menu bar
3. Next, select the **Delete Check** option below
4. Next, the system will provide you with a warning saying "are you sure you want to delete this transaction, etc.". If you are sure, click at the **OK** button and instantly the transaction will be completely removed from the system.

The Audit Trail

The **Audit Trail** is a report that's available in QuickBooks® and its purpose is to keep track of changes that occur in your system such as when you edit, void, or delete a transaction.
In order for the Audit Trail report to work, you have to set the Preferences in the program.

To set the Audit Trail **Preference,** select the following:
1. The **File** at the menu bar (in QuickBooks® 2000, select **Edit**)
2. The **Preferences** option below.
3. Next select the **Accounting** preference
4. Next, select the **Company Preferences** tab and click at the **Audit Trail** option.
5. Click at the **OK** button to save your settings.

To *view* or *print* the **Audit Trail** report, select the following:
1. The **Reports** at the menu bar
2. The **Other Reports** option (in QuickBooks® 2000, select **Accountant & Taxes**)
3. Next, select the **Audit Trail** option at the side menu

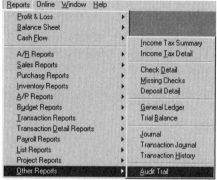

Figure 2.20 shows the location of the **Audit Trail** report.

Petty Cash

*P*urpose: To establish an amount of cash in the office to accommodate small purchases.

To start the **Petty Cash** fund, you will need the following:
1. An account in the chart-of-accounts that is *Other Current Asset* type.
2. A *vendor* with the name *Petty Cash* that you can use when you need to print a check to replenish the petty cash.

After you create the account and the vendor you will need to create a check that you can cash at the bank and bring the cash back in the office. That will start the Petty Cash fund.
To create a check for the Petty Cash fund, select the following:
1. The **Activities** at the menu bar.
2. The **Write Checks** option (in QuickBooks® 2000, select **Banking** and **Write Checks**).
3. At the **Pay to the Order of** field select the vendor named *Petty Cash*.
4. Next, click at the **To Be Printed** field so that the check will be printed.
5. Type the *amount* that you need to print the check at the **Amount** field.

6. Next, select the **Expense** option below, and from the chart-of-accounts select the *Petty Cash* account.
7. Type the amount at the **Amount** field. Let's say the amount that you need is $200. Through this transaction, the *Petty Cash* will increase by $200 via a *debit* type entry. The *Checking Account* will decrease by $200 via a *credit* type entry.
   The accounting that took place after this transaction was recorded, is the following:

	DR	CR
Petty Cash account	$200	
Checking account		200

8. Next, click at the **OK** button to create the check

Note: The cash that you bring from the bank after cashing the check should be locked-up in a box.

After you establish the Petty Cash, you can use the cash to buy various misc. supplies that you could use in the office. Remember however, to retain the receipts so that you can record the transactions properly and have the evidence to the purchases.

To record purchases made with money from the **Petty Cash** fund:
1. Select the **Lists** at the menu bar.
2. The **Chart-of-Accounts** option.
3. Next, select the **Petty Cash** account and click twice on it to open its *Register* and proceed to record transactions.
   For Example, lets say that you take a $10 bill from the **Petty Cash** box to purchase $5 worth of postage and $3 worth of office supplies. The total of the purchase is $8. Therefore, you should have a receipt of $8 and $2 change. Place both, the $2 change and the receipt into the Petty Cash box. Next, record the transaction.
4. Select the *date* under the **Date** field.
5. Next, type a *reference* under the **Ref** field that pertains to this transaction
6. Under the *payee* field, select the Vendor name from the Vendor list. For petty cash transactions, it maybe wise to create a new vendor with the name *Miscellaneous or Petty Cash* vendor so you don't accumulate all kinds of names in the system.
7. Next, under the **Decrease Account** column, type the amount of the transaction, in this case $8.
8. Next, because the example mentions that two things were purchased, postage and supplies, you need to select the **Splits** button as shown in figure 2.21. The Splits button allows you to record transactions that have more then one entry such as the current example. Click at the **Splits** button at the lower portion and you're ready to

record the two entries:
a. Under the account field, select the **Postage Expense** account, which is an expense account and type next to it the amount of the purchase for the postage, $5.
b. Next, select the **Office Supplies** account and type $3.
7. Next, click at the **Record** button to post the transaction.

Figure 2.21 shows the **Petty Cash** Register and the new transaction.

Notice in figure 2.21 that before the $8 transaction was recorded, the balance was $452. After the new transaction, the balance is $444 in the Petty Cash. The difference between the new balance ($444) and the previous ($452) is the $8 purchase, which is represented by the **receipt**. That means at any one time, the money that's left in the Petty Cash box, plus the total of the receipts should bring it to the amount that you started at the beginning of the month. And that is how you always must balance the Petty Cash fund.

Transfer of Funds

*P*urpose: To allow you to record transfers of funds made via telephone.

For example, if you make a transfer of funds from account A to account B at the bank using the telephone, you need to follow through by recording this transaction into the system.

To record a **transfer** of funds by phone, select the following:
1. The **Activities** at the menu bar (in QuickBooks® 2000, select **Banking**)
2. The **Transfer Money** option (in QuickBooks® 2000, select **Transfers Funds**).

78

3. At the next screen, and at the **Transfer Funds From** field, select the account you want to transfer funds from
4. At the **Transfer Funds To** field, select the account you want to transfer funds to.
5. Type the *amount* you want to transfer at the **Amount** field.
6. Next, make sure you have the right date in the date field and click the **OK** or **Next** button to *post* the transaction and you have finished recording the transfer of funds.

The accounting completed through this transaction, is the following.

	DR	CR
Checking Account A	$X	
Checking Account B		X

X = the amount of the transaction.

Figure 2.22 shows the Transfer of funds.

Purchase Orders

*P*urpose: To issue **Purchase Orders** for the purpose of ordering product and or services.

The Purchase Order function is a *non-posting* function; meaning when you create a Purchase Order it doesn't affect the data in your system. The Purchase Order function provides two benefits to your company:
1. When you create Purchase Orders, QuickBooks® creates a valuable report that's available to you. This report can help you manage your company's *cash flow* because it provides you with the information that is necessary in order to know future commitments that need to be paid.
2. When you use the Purchase Order function together with a simple policy that designates a person that is authorized to sign P.O's, it can help you eliminate *unauthorized* purchases, which often are found to be a problem in the small and mid-sized companies. Unauthorized purchases increase costs unnecessary.

To create a Purchase Order for **product**, select the following:
1. The **Activities** at the menu bar (in QuickBooks® 2000, select **Vendors**)
2. The **Create Purchase Orders** option from the drop down menu.
3. Next, select the *name* of the vendor from the Vendor list at the **Vendor** field.
4. Select a Class if you want to track the P.O into a Class, i.e. department. Next, you may want to track P.O's by Customer or Job name into a report so that you will have a record of P.O's by name. To track by Customer or Job, select a name at the **Ship To** field.
5. Next, you may set any of the misc. fields such as the:
   a. **Terms** according to your agreement with the vendor
   b. **Expected** date of delivery,
   c. **Ship Via** field. This information is all for historical purposes.
6. Next, select the product you wish to order by selecting an item from the Items list under the **Items** column
7. Next type the *quantity* you want to order at the **Qty** field for any particular item. QuickBooks® does the multiplication of the quantity and the rate (the rate is the cost of the product or the service. It comes defaulting when you select the Item).
   To add another Item on the form, repeat the process.
   Purchase Orders must be printed on paper so that an employee may sign them.
   Note: QuickBooks® prints P.O's on plain paper.

8. To finish creating a Purchase Order, click the **OK** or the **Next** button.

To create a Purchase Order for a **Service**, select the following.
Let's say you are a service type company and happen to be very busy and a good customer called requesting your immediate attention. However, you are working with another company-a subcontractor, on an as-needed basis and use them to provide the service to your customer.
So, you employ their services to provide the service to our customer and then upon completion, they bill you for this service. And of course you will follow-up by billing the customer for the service provided.
Therefore, according to company policy, we will have to fax a Purchase Order to the sub.
To create a P.O for a service, you must follow the same steps as in creating a P.O for a product. The only difference is the Item will be a **Service** type.

After you create the Purchase Orders, there are valuable reports that are available to you such as a List, Purchase Orders by customer and Open Purchase Orders.

A. To print the Purchase Order **list**, select the following:
   1. The **Lists** at the menu bar (in QuickBooks® 2000, select **Reports / Purchases** and the report).
   2. The **Purchase Orders** option from the drop down menu.
   3. Next, click at the **Purchase Orders** button at the bottom of the *Purchase Orders* screen as shown in figure 2.23
   4. Next, click at the **Print List** button.

Figure 2.23 shows the P.O list.

B. To print the **Open Purchase Orders** report, select the following:
   1. Select **Lists** at the menu bar (in QuickBooks® 2000, select **Reports / Purchases** and the report).

2. The **Purchase Orders** option from the drop down menu
3. Next, click at the **Reports** button to the right
4. Next, select the **Purchase Orders** option, and the **Open Purchase Orders** report.

C. To print the **Open Purchase Orders by Job** report, select the following:
1. Follow steps 1 through 3 above
2. Next, from the **Purchase Orders** option, select the **Open Purchase Orders by Job** report.

You should print the **Open Purchase Orders** report on a weekly basis so that you can be an informed manager about such issues as: when the delivery is due, how much money the purchase order is made out for, the name of the vendor, the date, etc. Anytime you issue a Purchase Order, the vendor will provide the product or service that you have ordered, and you must be prepared to pay for the service or product that has been delivered. So, you're not surprised when those deliveries are made.

To *print* the report, select the **Print** button on the upper section of the report.

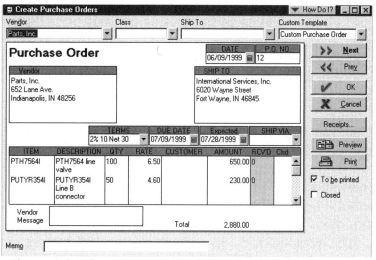

Figure 2.24 shows a completed *Purchase Order*.

Enter Bills

*P*urpose: To record the vendor bills (invoices) your company receives for purchases of product or services.

Vendor bills must be recorded through the **Enter Bills** function. When you record the vendor bills in the system, you complete the necessary accounting that is required and automatically, QuickBooks® updates the financial statements, which the manager then can use in order to make good sound business decisions.

Business tip: *Do not pay bills immediately after you have received them. Pay bills when they are due. This way you maximize your company's cash. Of course, the company's cash should be in an account, at the bank, that earns interest.*

Some of the vendor bills you receive may be connected to a Purchase Order and some others may not. Vendors such as the Telephone, Utility and the Electric companies (and others) do not accept a P.O. When you receive a bill, simply proceed to record it via the **Enter Bills** function. If the bill is from a vendor you have previously issued a P.O, QuickBooks® will inform you because it connects the Purchase Order with the vendor name.

After you have issued **Purchase Orders**, the vendors that you've ordered product or services will follow through with the delivery, and they will also send you a bill so they can get paid.

When you *record* a vendor bill, you update the P&L and the Balance Sheet statement automatically. Also, QuickBooks® creates the Account Payable (A/P) report.

Example #1: Let's say that you have received a bill from the *telephone* company for the month X service. The bill is for $400.
To record the **phone bill**, select the following:
1. The **Activities** at the menu bar (in QuickBooks® 2000, select **Vendors**).
2. Next, the **Enter Bills** function.
3. Click next at the **Vendor** field and select a name from the vendor list
4. Next, at the **Date** field, type or select the *date* from the calendar button
5. Next, type the bill's *reference* number (invoice number) at the **Ref. No** field
6. At the **Amount Due** field, type the *amount* of the bill (the amount of the purchase).
7. Next, you need to specify the accounting in the section below. An expense such as the

telephone service must be recorded into an *Expense* type account. To accomplish this task, click at the **Expenses** option, and select the account from the chart-of-account under the **Account** column. The account needed is an *Expense* type account such as the *Telephone* expense.

Note: Remember, in QuickBooks® you don't have to scroll to find an account (or for that matter a Vendor, a Customer or another name). All you have to do is type a few characters, and instantly you'll see the name of the account, vendor, customer, etc. highlighted on the list.

Now here again, we will split the amount of the bill (of $400) into two $200's and select the account twice (for each amount) as its shown in figure 2.25.

8. Next, click under the **Class** column and select a Class or department in order to track the amounts into a Class (or department) for departmentalized accounting purposes.
9. Click at the **OK** or the **Next** button to complete the recording and post the bill in the system.

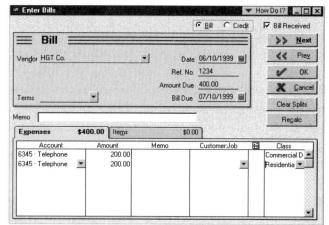

Figure 2.25 shows the recorded bill.

The accounting completed when you record a bill via the **Enter Bills** function is as follows: The amount that appears on the bill is a *credit* type entry and increases the *Accounts Payable* account. The amounts under the Expense option (as is the case in this example), as shown in figure 2.25, are a *debit* type entry and they increase the *expense* account.

	DR	CR
Telephone Expense	$200	
Telephone Expense	$200	
A/P account		400

If you intend to record more bills, click on the **Next** button, so you may have a new screen to record another, new transaction.

To record similar type of bills for your particular type of business, follow the same steps as in the example above.

Business tip: In order to be efficient, remember to accumulate a few bills at a time and record them all in one step. A good practice may be that you designate a specific day during the week for entering vendor bills.

Example #2: Let's say that you purchase materials on account (to pay later) from your supplier in order to use in a Job (project) that your company is currently working on for customer X.

To record this bill, follow all the steps from Example # 1 above, except in step # 7 select a different account such as the *Material Purchases* account (expense type account).

Job Costing Expenses
Because the materials were used in a job, as we have stated earlier, you need to track the cost of the material into the job report (for job costing purposes).
To do **Job Costing**, select the following:
1. Click under the **Customer:Job** column and
2. Select a *job name* from the *Customer:Job* list
3. Click at the **OK** or **Next** button to post the transaction into the job report

Note: To create new jobs for customers, please refer to the section about Entering Customers, in Chapter 1.
To complete this example, we need to emphasize the accounting that took place. Let's say that you have purchased material worth $850. When you have finished recording this bill via the **Enter Bills** function, the accounting in the background will be as follows:

	DR	CR
Material Purchases	$850	
A/P account		850

Example #3: A *homebuilder* that builds spec homes, makes purchases of material and supplies to complete various projects. When vendors bill you for these purchases, you must record the bills you receive via the **Enter Bills** function and into an asset account. In other words, you must <u>not</u> expense the cost of the purchase. If you do not have this account in the chart-of-accounts, you can create a new one. The account type must be **Other Current Asset**. This account can be named Work in Progress.

To record bills for material you have purchased to build a spec house, follow the steps outlined in Example #1 except in step # 7 select the *Work in Progress* account.

Let's say you've purchased lumber for $6,500 on account (pay later) that you will be using to complete project X. After you have finished recording the bill you've received from the vendor via the **Enter Bills** function, the accounting in the background will be as follows:

	DR	CR
Work in Progress	$6,500	
A/P		6,500

By recording the cost of the material into an asset account, you will not affect the P/L statement, only the Balance Sheet (that's where asset accounts are reported). Because you will not have these purchases on the P/L to monitor the progression of the costs, it is important that you do job costing so that you may be able to monitor the company's performance in each spec house via the job name report (from job costing).

Also, the **advances** you receive from the bank to complete the spec houses, must be recorded into a liability account that is **Other Current Liability** type (we'll be discussing the recording of advances at the *Make Journal Entries* section, in Chapter 5). Note: Please consult with your accountant before you begin the type of accounting described above.

When materials are used in a Job, you need to track the cost of the material into a job report (job costing) so that you can monitor the job profitability.

Job Costing Expenses
To do **Job Costing**, select the following:
1. Click under the **Customer:Job** column and
2. Select the *job name* from the Customer:Job list
3. Click at the **OK** or **Next** button to post the transaction into the job report

Converting P.O's into Bills

Example #4: Let's say that the vendor you have issued a **Purchase Order** at an earlier time, has delivered the product and has also sent a bill.

To record a bill that is connected to a **Purchase Order** after the product has been delivered or a service is provided, select the following (just as in example # 1):
1. The **Activities** at the menu bar (in QuickBooks® 2000, select **Vendors**).
2. Next, the **Enter Bills** function.
3. At the **Vendor field** select the vendor name from the vendor list.
   The minute you select the name, and click anywhere on the screen, QuickBooks® opens a reminder saying, " Open Purchase Order exists for this vendor. Do you want to receive against one or more of these Orders?" as shown in figure 2.26.

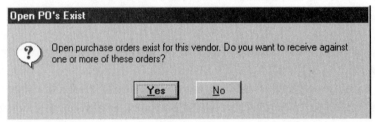

Figure 2.26 shows the "Open P.O's Exist" screen.

In this window, there are two options: **No** and **Yes**:
a. The **No** option allows you to record another bill for the same Vendor. Unrelated to the Purchase Order.
b. The **Yes** option allows you to <u>receive</u> the Purchase Order.

To **receive** a P.O,
- Click at the **Yes** button and at the next screen
- Click under the √ column to place a checkmark (√)
- Click at the **OK** button to complete the <u>receiving</u> of the Purchase Order.
  Instantly the Purchase Order turns into a bill.

Because you have received a P.O that was issued to order product, the **Items** option was automatically selected in the **Enter Bills** function, as its shown in figure 2.27. These Items are the same Items that were selected to create the P.O. Through the Items, the product that was purchased will be directed in the

Inventory account (that's the default account used by *Inventory Part* type Items).
4. Next, at the Bill screen, select the *date* at the **Date** field to be the same as the bill's date
5. Type in the bill's *reference* number at the **Ref. No** field
6. Next, click at the **OK** or the **Next** button to complete the recording of the bill.

If you receive a bill that is *different* from the Purchase Order you can adjust the quantity and the amounts on the bill. For example, let's say that the bill has an extra charge for freight and also the quantity of an Item is different. To adjust for such differences, follow the steps below:
1. To record for example a **freight** charge:
   a. Select the **Expense** option and from the chart-of-accounts
   b. Select an account such as *Shipping* or *Freight* expense. Type the amount of the freight charge, let's say $70, and click anywhere on the screen. Instantly you've recorded the expense of the freight in the same step with the receiving of the P.O.
2. To record a difference in the **quantity** of the product received (when the quantity on the bill is not the same as it was on the P.O), just type the new quantity over the previous quantity in the **Qty** field. For example, let's say the vendor had called and said that they couldn't deliver all 100 pieces of a particular product we had requested via the P.O and that they could only ship 90. All you have to do is just change the bill by typing the new quantity, in this case 90, in the **Qty** field.
1. Next, click the **OK** or the **Next** button to complete recording the transaction.

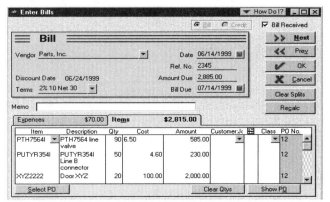

Figure 2.27 shows both, the expense of $70 (at the Expenses tab) and the receipt of the items for $2815. This is the completed purchase transaction.

The accounting completed in the background from this transaction, is the following:

	DR	CR
Shipping expense	$70	
Inventory account	$2815	
A/P account		2885

To convert Purchase Orders into bills for your particular type of business, follow the same steps that were outlined in the example above.

If you <u>receive</u> a **partial** P.O, as was the case with this example (you had ordered quantity of 100 in Purchase Order but had received only 90), the system remembers the difference and keeps the Purchase Order open.

To record the receiving of the balance of the P.O after the vendor delivers the product and send you the bill, simply follow **all** the steps outlined above during the receipt of the original P.O. After you place a checkmark (√) next to the P.O and click the **OK** button, QuickBooks® brings the balance of the undelivered items from the P.O into the new bill. <u>And that's how you manage back orders</u>.

The accounting that occurs from this transaction is the same as in the previous example.

When a Purchase Order is received in <u>full</u>, it is removed automatically from the **Purchase Orders** list as shown in figure 2.28 (there used to be two P.O's).

Figure 2.28 shows the P.O that is left on the Purchase Orders screen.

Closing a Purchase Order

If you have a Purchase Order in the P.O list has been *partially* received and you don't need the balance of the product or the service listed on the P.O, you have two options: You can *delete* it or *close* it.
1. If you *delete* a P.O it will be removed from the system.
2. If you *close* a P.O it will remain in your history file, and later you may view or print.

1. To **Delete** the P.O, follow these steps:
   a. Select the **Lists** at the menu bar (in QuickBooks® 2000, select **Reports** and **Purchases**).
   b. The **Purchase Orders** option (in QuickBooks® 2000, select **Open Purchase Orders**).
   c. *Highlight* the P.O on the list that you want to delete (click twice on the P.O in v.2000).
   e. Select the **Edit** at the menu bar.
   e. Select the **Delete Purchase Order** option.
   f. Click at the **OK** button.
      The Delete function will delete permanently the P.O from your records. That means you lose the history and that's not a preferred option.
2. To **Close** the P.O, select the following:
   a. The **Lists** at the menu bar (in QuickBooks® 2000, select **Vendors**).
   b. The **Purchase Orders** option (in QuickBooks® 2000, select **Purchase Order List**).
   c. *Highlight* the P.O on the **Purchase Orders** list and click twice on it to open it.
   d. Next, click at the **Closed** field to the right as shown in figure 2.29
   e. Click at the **OK** button to save your selections.
   f. Next, select the **Yes** option in the next screen.

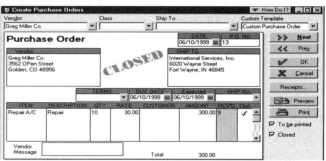

Figure 2.29 shows a closed **Purchase Order**.

Closing the P.O. is the preferred option because it is part of the history.

To *view* or *print* a closed Purchase Order, select the following:
1. The **Lists** at the menu bar (in QuickBooks® 2000, select **Vendors** and **Purchase Order List**)
2. The **Purchase Orders** option at the drop down menu
3. Click at the **Purchase Orders** button
4. Next, click once at the **Show All Purchase Orders** option and you will find all the P.O's you have issued in the past. That's the P.O history.

A closed P.O can be reprinted in case of a dispute. To reprint a P.O from the history file, select the one you want to print and click twice on it to open it. Next, click at the **Print** button to the right to print.

A/P report

The bills you record in QuickBooks®, automatically are entered in the **Accounts Payable (A/P)** report whose purpose it to provide you with detail information of the various amounts your company owes to vendors and suppliers. These are the companies that you do business with by purchasing their product and services.

To print the **A/P** report, select the following:
1. The **Reports** at the menu bar.
2. Next, select the **A/P Reports** option (in QuickBooks® 2000, select **Vendors & Payables**)
3. Next, select either the **Aging Summary** or the **Aging Detail** report.

The *Detail* is the preferred report because it shows every transaction separately. It is one of the reports you need to print on a *weekly* basis alone with the P/L and the Balance sheet. Its purpose is to show you the amounts your company owes to each vendor, the date that it's due, and the aging of each amount.

Credit Card Charges & Credits

𝒫urpose: To record purchases (and credits) made using Credit Cards.

QuickBooks® has the capability to manage credit card type transactions such as *purchases*, *credits*, *finance charges,* and payments to the credit cards.

To be able to work with credit cards properly, you must create an account in the chart-of-accounts first for every credit card you intend to use for business purposes.

To create an account, follow the steps we have outlined and the *About The Chart-of-Accounts* section, in chapter # 1. The type of the account for a credit card must be **Credit Card**.
The Credit Card type account is a liability account and will post in the *Current Liabilities* section on the Balance Sheet.

At the bottom of the **New Account** window, during the creation of the new account, as shown in figure 2.31, there are two fields, the **Opening Balance** and the **as of** field:
   1. The **Opening Balance** field allows you to enter a *beginning balance* for the new credit card account
   2. The **as of** window is where you can type a *date* for the beginning balance.

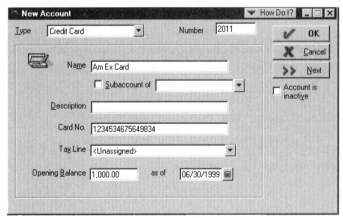

Figure 2.31 shows a *new account* for a credit card.

If the credit card (for which you're creating the account) is new, all you have to do is select:

1. The *Type* of the account
2. Type a *name*
3. Enter the credit card's *number*
4. Click the **OK** button to save the information for the new account.

However, if the credit card is not new but its an existing credit card, then most probably it would have an outstanding amount that is due at the time you are creating the new account in chart-of-accounts. This outstanding amount you can enter at the time when you create the new account.

For example, let's say you have an existing credit card and as of the end of last month X it had an outstanding balance due of $1,000. You want to create an account for this credit card in QuickBooks®. After you enter the information for this new account as outlined in chapter # 1 and above, than you can type the outstanding amount of $1000 at the **Opening Balance** field as shown in figure 2.31. Next, select a date at the **as of** field. Figure 2.31 above, shows the new credit card account with the opening balance entered.

When you type the amount of the opening balance, QuickBooks® automatically completes the transaction for you with an offsetting entry. This offsetting entry goes into an account that's called **Opening Balance Equity**. This account is an Equity type account.

Note: The *Opening Balance Equity* account is used by QuickBooks® to offset single entry transactions such as the one described in this example. *After you enter the opening balance, you must follow through with an adjusting transaction via the <u>Make Journal Entries</u> function and reverse-out the amount from the Opening Balance Equity account. After you enter an outstanding balance, the Opening Balance Equity account carries a debit type entry (as shown below). To reverse it you must enter a credit type entry. Please, consult with your Accountant about the account you need to use for the offsetting entry to this adjustment.*

The accounting that takes place in the background for the credit card *opening balance* amount is as follows:

	DR	CR
Opening Bal. Equity	$1000	
Credit Card account		1000

Note: The Credit Card account is an account in the chart-of-accounts.

Once you create the credit card account in the system, you are ready to record transactions.

Any time you make a purchase using a Credit Card, you need to retain the **receipts** that you receive from the vendors and record purchases in the system *through these receipts*.

Business tip: *Do not record purchases in the system from the credit card statement you receive at the end of the month because the statement may contain fraudulent and faulty transactions.*

It is recommended that you record purchases made with credit cards once or perhaps twice a week depending on the volume of your business, at designated days such as a Wednesday or Friday. So than when you print the financial statements at the beginning of the week, they will also include the credit card transactions.

Example #1: Purchased $145 of office supplies using the company's Credit Card from Office Supplies One (a vendor).

To record this purchase made with the **credit card**, select the following:

1. The **Activities** at the menu bar (in QuickBooks® 2000, select **Banking**)
2. The **Enter Credit Card Charges** option from the drop down menu
3. Once you are in the **Enter Credit Charges** window, select Master Card from the credit card list
4. In the **Purchased From** field, select the *Office Supplies One* name from the vendor list
5. Next, type the receipt's *reference* number at the **Ref No.** field
6. Type or select the *date* of the purchase at the **Date** field
7. Next, select the **Charge** option to record the purchase. Next to this field is the **Credit** option. Here is an explanation of the two options:
   a. The *Charge* field is to record *purchases*
   b. The *Credit* field is for *credits* you may get when you return merchandise.
8. Next, type the *amount* of the purchase at the **Amount** field.
9. Next, at the lower part of the screen, select the **Expenses** tab option to expense this purchase and click again under the **Account** column, select an expense account that is proper for this purchase such as the Office Supplies account.
10. Next, skip over the **Customer:Job** field because job costing is not applicable
11. At the **Icon** field, click once on the Icon to make it non-billable (as discussed earlier)
12. Next, you can select a **Class** to track the expense into a Class (or department) for departmentalized accounting purposes if that is applicable
13. Next, click at the **OK** or the **Next** button to post the purchase
    The accounting that was completed in the background by recording this purchase

made using a Credit Card, is as follows:

	DR	CR
Office Supplies account	$145	
Master Card account		145

Example #2: Using the company's <u>Master Card</u> you make a purchase $760 worth of material from Supplier X to use at a Job (project) your company is currently working on a T&M basis.

To record this **credit card** purchase, select the following:
1. The **Activities** at the menu bar (in QuickBooks® 2000, select **Banking**)
2. The **Enter Credit Card Charges** option from the drop down menu
3. Once you are in the **Enter Credit Charges** window, select the credit card you need to work with from the list
4. In the **Purchased From** field, select the Supplier X name from the vendor list
5. Next, type the receipt's *reference* number at the **Ref No.** field
6. Type or select the *date* of the purchase at the **Date** field
7. Next, select the **Charge** field
8. Next, type the *amount* of the purchase at the **Amount** field.
9. Next, click at the **Expenses** tab to expense this purchase and under the **Account** column, click and select the Material Purchases expense account.
Job Costing Expenses
10. <u>Next, at the **Customer:Job** field select the job *name* where the material was used in order to track the cost of the material into the job report (for job costing).</u>
11. <u>At the **Icon** field next, leave the Icon intact (do not do anything) in order to make this expense a billable expense. If it is not billable, you need to click once on the Icon (as discussed earlier) to remove the amount from the billable list.</u>
12. Next, you can select a **Class** to track the expense into a Class (or department) for departmentalized accounting purposes.
13. Next, click at the **OK** or the **Next** button to post the purchase.
    The accounting that was completed from this purchase that was made using the credit card is as follows:

	DR	CR
Material Purchases account	$760	
Credit Card account		760

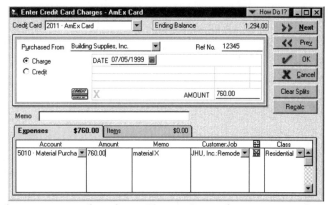

Figure 2.32 shows a completed **credit card** purchase transaction.

If you need to record the purchase of product that you want to place in inventory for later use or to sell, follow all the steps outlined in **Example #2** except, in step # 9 select the **Items** tab instead of the Expenses tab.

To record a credit card **Credit** transaction for merchandise you have returned to the vendor, follow all the steps outlined in the above examples except, in step # 7 select the **Credit** option instead of the **Charge** option. A credit card **Credit** must be recorded using the same expense account that was used originally to record the purchase.

Finance charges from credit cards can be recorded as follows:
1. As another **Charge** type transaction like the examples described above or
2. Through the **Reconcile** function which we will discuss in the next section.

Printing the **Balance Sheet** report will help you monitor the amounts your company owes to Credit Cards. These amounts are posted in the *current liability* section. The Balance Sheet should be printed on a weekly basis.

International Services, Inc.
Balance Sheet
As of July 31, 1999

	Jul 31, '99
Accounts Payable	
2000 · Accounts Payable	2,800.00
Total Accounts Payable	2,800.00
Credit Cards	
2011 · AmEx Card	2,054.00
2005 · VISA BankCard Inc.	130.00
2007 · MC Bank Inc.	3,036.01
2009 · Discover Card	1,874.00
Total Credit Cards	7,094.01

Figure 2.33 shows the Balance Sheet section of the *current liabilities*

Reconciling the Credit Card Statement

*P*urpose: To ensure that there are no errors or fraudulent transactions on the Credit Card statement.

At the end of each month, you receive the statement from the credit card that shows the amount your company owes them. Before you make a payment, you must first **reconcile** the statement against the information you have accumulated in the system by recording the vendor receipts.

The process of reconciling the credit card statement against the system's information is accomplished through the **Reconcile** function. This function is also used to reconcile Bank account statements.

The <u>objective</u> of the reconciliation process is that you end-up with a zero (0) difference. That will indicate that the *amount* shown on the statement as owed to the credit card, agrees with the information you have accumulated in the system (or equals the transactions you have recorded).

To **reconcile** the credit card, follow these steps:
1. Select the **Activities** at the menu bar (in QuickBooks® 2000, select **Banking**).
2. The **Reconcile** option from the drop down menu
3. At the **Account To Reconcile** field in the *Reconcile Credit Card* window, select the Credit Card **name** from the list that you want to reconcile
4. Next, type the *ending balance* amount that appears on the statement at the **Ending Balance** field
5. Next, type in the *finance charge* amount that appears on the statement, in the **Finance Charges** field. Let's say the amount is $10
6. Next, select the *date* of the finance charge
7. At the **Account** field select the *Interest Expense* account to record the finance charges.
   After typing the finance charges the accounting that takes place is as follows:

	DR	CR
Interest Expense account	$10	
Credit card account		10

8. Next, begin the process of reconciling.
   In the Reconcile screen, there are two sections: **Charges and Cash Advances** (the lower section) and the **Payments and Credits** section (upper section). Their names clearly indicate what each contains. You may elect to start with either one.
   To begin the reconciliation process, follow the steps below:
   a. View the following fields and make sure that what's in the system agrees with what is on the statement: **Date, Ref. No., Payee, and Amount**.
   b. If the information in the above fields *does* agree on both, the statement and the system, then place a checkmark (√) next to the transaction by clicking once.
      Placing a *checkmark* next to a transaction, means the transaction is OK and it also tells QuickBooks® that the transaction has **cleared**.
   c. If the transaction appears on both, the statement and in your system, but it doesn't agree i.e. the amount is different. Then you need to find the receipt and make sure it was recorded properly. If, from your end it is OK, then you need to contact the credit card issuer and report the error.
   d. If a transaction appears in your system, but it doesn't appear on the statement, that indicates that the transaction did not *clear* the credit card's process. Do not place a checkmark (√) next to it. The absence of a checkmark tells QuickBooks® that the transaction has *not* been *cleared* yet.
      Note: Transactions that haven't cleared (that you did not place a checkmark) will

appear again during the next month's reconciliation. On the other hand, transactions that have been cleared (those with a checkmark) will be removed from the Reconcile function once you finish the process and click the DONE button.

e. If you end-up with a difference (not zero), it could mean one of three possible type of scenarios:

Type	Action
An error	Correct the entry if it's your error and proceed to complete the reconciliation. If not, contact the credit card Company and report the difference. Do not finish the reconciliation. Wait for the correct amount from the credit card Company to adjust your entry.
Fraud	Contact the credit card company ASAP to report the fraudulent transaction. Do not finish the reconciliation. Wait for the correct amount from the credit card Company to adjust your entry.
No record	Using the vendor receipt, a. Record the transaction via the **Enter Credit Card Charges** function b. Return to the **Reconcile** function to complete the process.

*Please remember: If you end up with a difference, do not click the DONE button. You must click the **Leave** button. If you click on the Leave button, QuickBooks® will remember all the work you have done so that you do not have to repeat the task of reconciling again. Later, when you return to the Reconcile function, you can continue where you had left off.*

f. In the event of a *credit card error* or *fraud*, you need to contact the credit card Company as soon as possible. After you resolve your difference with the credit card Company, you should receive an adjusted statement with the correct balance. At that time, you need to return to the **Reconcile** function again and adjust the Ending Balance figure and continue the reconciliation process where it was left off. After the adjustment, you should have a zero (0) difference as shown in figure 2.34 below.

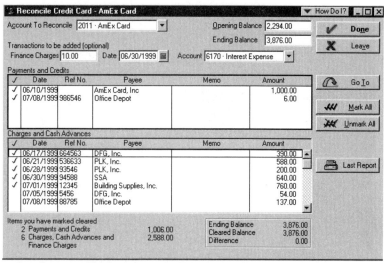

Figure 2.34 shows a completed reconciliation process. **Reconcile Credit Card** screen.

g. When you have finished the reconcile process and ended up with a zero (0) difference, then click at the **DONE** button (upper right corner).
h. After you click the **DONE** button, the **Make Payment** screen appears. In this screen you have two options. You may select to:
- **Print** a *check* to pay the credit card as shown in figure 2.35.

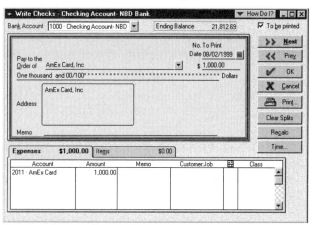

Figure 2.35 shows the **check** to pay the credit card (**Write Checks** function).

100

If you select to print a check for the credit card, here is the accounting that takes place (assuming a payment of $1000):

	DR	CR
Credit Card account	$1000	
Checking account		1000

- **Create** a *bill* as shown in figure 2.36. When you choose to create a bill, QuickBooks® enters a new bill in A/P via the **Enter Bills** function. This bill can <u>only</u> be paid through the **Pay Bills** function, which we will discuss in the next section.

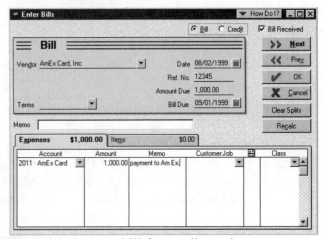

Figure 2.36 shows a bill for credit card payment

The accounting completed by creating the *bill* is as follows (assuming a bill for $1000):

	DR	CR
Credit Card account	$1000	
A/P		1000

From either function, you may pay a *partial* payment by simply changing the amount under the **Expense** column, at the lower section of the *check* or *bill* screen.

Note: The amount in the **Opening Balance** field in the *Reconcile Credit Card* screen can only be changed via the account's **Register**. If you change any previously recorded and cleared (reconciled) transaction in the register, it will affect the **Opening Balance** field in the *Reconcile Credit Card* screen.

Pay Bills

𝒫urpose: To create checks that are needed to pay the unpaid (or open) vendor bills.

Unpaid vendor bills can <u>only</u> be paid through the **Pay Bills** function.
Before you proceed with the process of creating checks of the unpaid bills, you should print the **A/P Aging Detail** report and mark with a pencil the bills you want to pay.

To print the **A/P Aging Detail** report, select the following:
1. The **Reports** at the menu bar (in QuickBooks® 2000, select **Reports**)
2. The **A/P Reports** option (in QuickBooks® 2000, select **Vendors & Payables**)
3. And the **A/P Aging Detail** option

Purchase Discounts

The **Pay Bills** function is where you can take ***purchase discounts***. A purchase discount is a percent from the bill's total that you get to keep provided that you pay the bill within a designated amount of days from the date that appears on the bill.

To create checks for the unpaid bills, select the following:
1. The **Activities** at the menu bar (in QuickBooks® 2000, select **Vendors**).
2. Select the **Pay Bills** function below.
3. At the Pay Bills screen, the **Payment Date** field is the date that goes on the checks. You may leave it as is. It is the system default date.
4. Click at the **Check** field indicating that you will pay with checks.
5. Next, you may select either the **Show bills due on or before** or **Show all bills:**
   a. The **Show bills due on or before** allows for a date cut-off
   b. The **Show all bills** shows all the bills that are unpaid as of the date you enter the Pay Bills function.
6. Next, you may optionally *sort* via the four options at the **Sort Bills By** field. Select the option by which you wish to have your bills sorted.
7. Next, select the *Checking Account* (in the event that you have more then one) that the amount to pay the bills would come from as shown in figure 2.37.

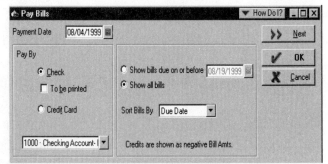

Figure 2.37 shows the *upper* section of the **Pay Bills** function.

8. Next, in the lower section of the screen you should be able to see all the unpaid bills. Proceed as follows:
   a. First, take any available **purchase discounts**:
      - Click under the √ field, next to a bill that has a date under the **Disc. Date** field (QuickBooks® creates a check for all bills that have a checkmark (√) next to them).
      - Next, click at the **Discount Info** button as shown in figure 2.38

Figure 2.38 shows the *lower* section of the **Pay Bills** function.

- At the **Discount Information** screen, the system will display all the information regarding the bill you have checked. Select an account at the **Discount Account** field that is *Income* type from the chart-of-accounts. This account becomes the default and will be used by the system to automatically record the *purchase discount* from that point forward.

103

- To take the discount, simply click at the **OK** button as shown in figure 2.39.

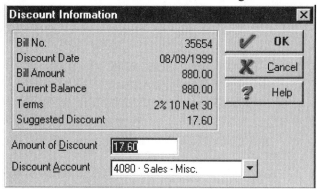

Figure 2.39 shows the **Discount Information** screen.

The accounting that has occurred by taking the purchase discount is as follows:

	DR	CR
A/P account	$17.6	
Sales – Misc. Income		17.6

b. Next, proceed by placing a checkmark (✓) next to each bill to create a check for it, as shown in figure 2.40. Notice the bill with the discount date that the amount under the **Amt. Due** field, is different then the amount in the **Amt. Paid** field. Also, notice bill # 2341, which had an amount due of $2400, but the amount under the **Amt. Paid** field, is $1500. That's a partial payment. To make a **partial payment**, highlight the amount under the **Amt. Paid** field and *type* over a new amount.

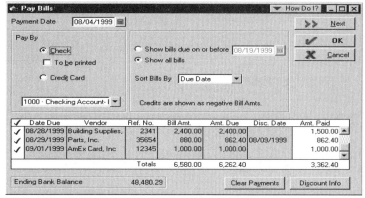

Figure 2.40 shows the completed **Pay Bills** screen with the selected bills to be paid.

104

c. Next, click the **OK** button and QuickBooks® will create the checks. These checks will be directed to the *print batch* file.
d. To **print** the checks, select the following:
   - The **File** at the menu bar
   - The **Print Forms** option from the drop down menu.
   - Click next to each check that you want to print by placing a checkmark (✓)
   - Click the **OK** button to print all selected checks as shown in figure 2.41.

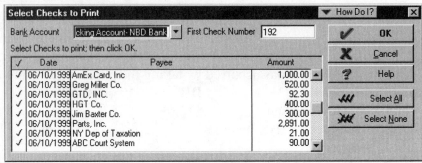

Figure 2.41 shows the selected checks that need to be **printed** from the *print batch*.

Using the Register

*P*urpose: To record business transactions.

In Chapter 1, under the *About the Chart-of-Accounts* section, we discussed the purpose of the **Register**. In this section, we'll explain how to record transactions by using the Register of an account.

To use the **Register** for recording transactions, select the following:
1. The **Lists** at the menu bar.
2. The **Chart-of-Accounts** from the drop down menu.
3. Select the account that you want to **record** the transaction by *highlighting* it.
4. Next click twice on it to open its **Register**. In the account's Register,

a. Select *date* of the transaction at the **Date** column
b. Type a transaction or reference *number* under the **Number** column
c. Select a *name* from the Vendor list
d. Type an *amount* under the **Payment** column
e. To complete the accounting, click at the button next to the **Account** field if the transaction has <u>one</u> entry.
   If the transaction has <u>two</u> or more entries, click at the **Splits** button below and type the amount next to the account that you select from the chart-of-accounts list
f. Next, click at the **Record** button to post the transaction.

Example #1: Recording a new *bank loan* (moneys your company borrowed from the bank) by using the *Checking* account's **Register**.

Figure 2.42 shows the recording of a new Bank loan (with one entry).

Example #2: Recording a *manual check* by using the *Checking* account's **Register**.

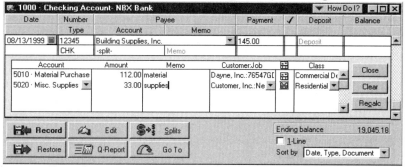

Figure 2.43 shows the recording of a Manual check, with two entries, with *Job costing*

and *departmentalized accounting* (via the Class field). Multiple transactions are recorded by using the **Splits** button at the lower section of the screen.

Example #3: Recording a *Vendor Bill* through the *Accounts Payable* account's **Register**.

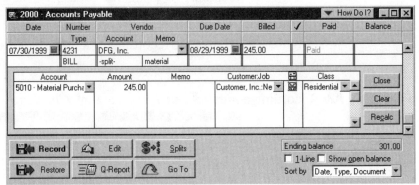

Figure 2.44 shows the recording of a Bill (with one entry).

Note: To track a purchase expense into a Job name (for job costing), you must use the Splits button as shown in figure 2.45 below.

Figure 2.45 shows the same transaction as in figure 2.44 (Example #3) but with **Job Costing** and **Departmentalize** accounting completed via the **Splits** button.

Pay Sales Tax

*P*urpose: To create checks for the sales tax that is due to the State or Local government.

In QuickBooks®, when you create a customer invoice the sales tax accumulates automatically in a *current liability* type account called *Sales tax payable*.
Depending on the amount of the sales tax, the State government determines how often the sales tax ought to be paid. Most companies however, seem to be on a monthly schedule. Companies that are on a monthly schedule, must pay the sales tax every month and before the end of the following month.
Please note: To find out about the time table of paying your sales tax, consult with your accountant or call the appropriate department of your State.

To create **checks** for the sales tax, follow these steps:
1. Select the **Activities** at the menu bar (in QuickBooks® 2000, select **Vendors**).
2. Select the **Pay Sales Tax** option from the drop down menu.
3. Select the *account* (where the money's coming from) at the **Pay From Account** field.
4. Next, make sure that the **To be printed** field is checked so that you may print the check.
5. In the **Pay** column, click next to the sales tax that is due to be paid to place a *checkmark*. That will create a check.
6. Next, click the **OK** button to create a check for the sales tax as shown in figure 2.46 below.
   The accounting completed for the payment of the sales tax, is as follows:

	DR	CR
Sales tax due (a liability account)	$55	
Checking account		55

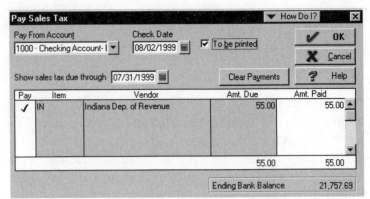

Figure 2.46 shows the **Pay Sales Tax** screen.

Sales Tax Credit

If your State government allows you to retain a *percent* of the sales tax amount that is due, known as *sales tax credit*, you can record it properly in QuickBooks® by following the steps bellow:

A. Create the check for the State government:
1. Follow steps 1-5 from the **create checks** process, above, to create the check.
2. Next, highlight the amount to be paid under the **Amt. Paid** field, and type over the new amount (just like a partial payment), which is, the original amount *minus* the credit amount you get to keep.
3. Click the **OK** button to create the check.

B. Take the credit:
1. Follow steps 1-3 from step **A** above.
2. Remove the (√) from the **To be Printed** field.
3. Click under the **Pay** column to place a (√) to create the check "to pay" and remove the amount left from this screen.
4. Click at the **OK** button.
   This process will remove any amount left in the *Sales Tax Payable* liability account and will empty the **Pay Sales Tax** screen

C. Adjust the credit:

1. Select the **Activities** at the menu bar (in QuickBooks® 2000, select **Company**).
2. Select the **Make Journal Entry** function and record the following adjusting transaction (let's assume the credit amount is for $5) as shown below and on figure 2.47:

	DR	CR
Checking account	$5	
Other Income account		5

You have finished recording the payment of the sales tax and the sales tax credit.

Business tip: The Other Income account that appears in the adjustment transaction must be an Other Income type account.

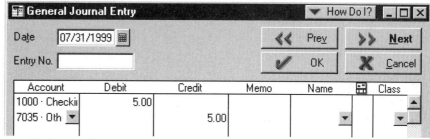

Figure 2.47 shows the **adjustment** entry for the sales tax credit.

Form 1099

It is the form you need to provide to non-employees at the end of each calendar year. Non-employees are people and companies that work for your company and fall into the following criteria:
    a. They conduct business as a proprietorship or partnership (they are not Incorporated)
    b. They have earned the minimum of $600 during the calendar year working for you
    c. They utilize their own resources such as equipment, office, etc.

To be able to print the 1099 forms, you must prepare your system first.

To prepare QuickBooks® to print 1099 forms, follow the steps below:
1. Prepare the vendor *template* in the vendor list:
    a. Select the **Lists** at the menu bar.
    b. The **Vendors List.**
    c. Highlight the vendor and click twice on it
    d. Select the **Additional Info** tab
    e. Click at the **Vendor eligible for 1099** field, and click at the **OK** button.
2. Set the *Preferences*:
    a. Select the **File** at the menu bar (in QuickBooks® 2000, select **Edit**).
    b. The **Preferences** option from the drop down menu.
    c. Scroll to the **Tax 1099** option and in the Company Preferences tab, select:
        - The YES option
        - At the **Box 7**, click under the **Account** column and select the expense account (s) from the chart-of-accounts list that you have been using to record the subcontractor costs. Next to this field and under the **Threshold** column, make sure the amount is set at $600.
        - Click at the **OK** button to save your preferences.

To print the **1099** forms, select the following:
1. The **File** at the menu bar
2. The **Print Forms** option at the drop down menu
3. The **Print 1099s** option from the side menu
4. Specify a **date range** by selecting the proper year-end
5. Click the **OK** button
6. Click next to each Vendor name you want to print a 1099 form
7. Click at the **Preview** button to *preview* each form

Click the **Print** button, at the upper left corner, to print the forms

Chapter 3

Income & Customer Payments

*T*he purpose of this chapter is to help you learn how to record properly all the transactions that pertain to the *income* your company will earn and the *payments* it will receive from customers.

In this chapter we'll examine the following QuickBooks® functions:

- **Create Estimates**
- **Customer Prepayments**
- **Make Deposits**
- **Create Invoices**
- **Receive Payments**
- **Enter Cash Sales**
- **Create Credit Memos/Refunds**
- **Enter Statement Charges**
- **Assess Finance Charges**
- **Create Statements**
- **Customer Bad Checks**

Create Estimates

*P*urpose: To create estimates that you can send to customers quoting the cost of your product or services.

The advantages of using the **Create Estimates** function are the following:
1. You can have professional looking estimates on plain white paper
2. Estimates will convert into invoices.
3. Estimates can help you do Progress Billing

When you create an estimate in QuickBooks® you doesn't affect the data in your general ledger because the **Create Estimates** function is non-posting. In other words, the amounts from the Estimate do not go anywhere.

To create an **Estimate**, select the following:
1. The **Activities** at the menu bar (in QuickBooks® 2000, select **Customers**).
2. Select the **Create Estimates** function at the drop down menu
3. In the next screen, select a Customer or Job name at the **Customer:Job** field. If you select a Job name in the estimate, the figures from the estimate's **Total** column will show-up IN the *Job Estimates* report. This report will allow you to compare estimated vs. actual figures (from the actual work that your company will do)
4. Select a *class* or department name at the **Class** field, to track the estimate by department
5. At the **Item** column, select by clicking from the Items list the appropriate Items you may need to complete this estimate. The Items that you can use could be Inventory Part, Service, Discount, Subtotal, Sales Tax, etc. like in figure 3.1 below.
   Let's say, for this example, that your company provides Heating & Air Conditioning services and a customer (or a prospect) asks you to provide an estimate to install a hypothetical commercial air-conditioning unit. The estimate must include labor and material.
   To complete this estimate you'll need the following Items from the Items list:
   a. Two Inventory Part type
   b. One Subtotal type
   c. One Discount type
   d. One Service type
   e. One Sales Tax type

6. When you have finished with the new Estimate, click at the **Print** button to print the estimate or simply click at the **OK** or the **Next** button to continue with creating another estimate.

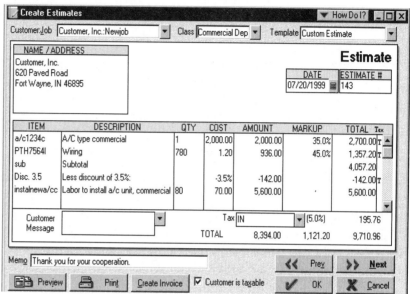

Figure 3.1 shows a completed **Estimate** with the selected Items.

Please notice the *markup* under the **Markup** column in figure 3.1. When QuickBooks® detects a *cost* and a *sale* number in an Item that is used on the Estimate, it instantly figures the markup under the **Markup** column. You can change the markup figure either up or down by highlighting.

On a completed Estimate there maybe information under the **markup** and **cost** columns like the ones on figure 3.1, that you may not want to share with your customer. To remove information that you don't want to show on a printed Estimate (or from other forms such as Invoices, Credit Memos, Enter Cash Sales, etc.), you can do it by *hiding* the columns that you don't want to appear on the printed copy.

To **hide** certain columns from a form, select the following:
1. The **Lists** at the menu bar.
2. Next, the **Templates** option from the drop down menu.

114

3. Select by highlighting the template you want to change. In this case the **Custom Estimate** template as shown in figure 3.2.

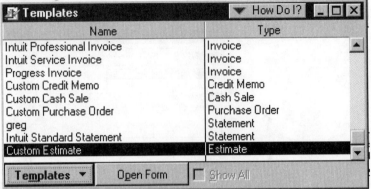

Figure 3.2 shows the **Templates** screen.

4. Click twice on the template you have highlighted (or select the Templates button below)
5. Select the **Columns** tab by clicking
6. Next, under the **Print** column remove any checkmarks (_) from the item you do not want to see printed on the paper copy of the form. Any item without a *checkmark* will not be printed on the printed document. On figure 3.3, there are only two columns left on the printed copy in comparison to the seven columns that were on figure 3.1.
7. Click on the **OK** button when finished to save the changes.

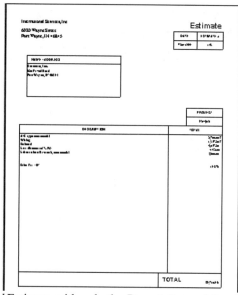

Figure 3.3 shows the *printed* Estimate with only the *Description* and *Totals* columns on it.

Customer Prepayments
(Customer Deposits or Advanced Payments)

*P*urpose: To record customer **prepayments** (deposits) that are received for services or product that you will be providing in the future.

Business tip: Customer prepayments are funds received from customers in advance. Before any service or product is provided. Because the product or the service has not been provided, income has not been earned and therefore, these types of funds must be classified as <u>unearned income</u>.

To record prepayments properly, you must first create an account in the chart-of-accounts and name it <u>Unearned Income</u>. This account must be an **Other Current Liability** type account. When you receive customer prepayments, they <u>must be recorded</u> into this liability account, through the *Enter Cash Sales* function and the use of an *Item*.

If your company is in a *construction* type business and requires a down payment from the customer before you begin work, that's a Customer Prepayment. When you receive the prepayment you must record it into the *Unearned Income* account.
If you are in the *retail* business and receive *layaways*, they are the same as prepayments.
If you are an *attorney* and receive *retainers*, they too are the same as prepayments.

No matter what type or size of business you may be in, you must record prepayments as a liability in your company books. If you don't, you will be doing the wrong type of *accounting* and may also *prepay* taxes.

To record a customer **Prepayment** that you've received, select the following:
1. The **Activities** at the menu bar (in QuickBooks® 2000, select **Customers**).
2. Next, select the **Enter Case Sales** function from the drop down menu.
3. In the **Customer:Job** field select the Customer or Job name from the customer list
4. Select the *date* in the **Date** field.
5. In the **Payment Method** field indicate the method of payment such as check, cash, Credit card, etc. Most likely this type of payment will come in the form of a check.
6. In the **Check No** field, type the check number (its important to have good customer history).
7. Next, in the **Item** column, select the item from the list that's called *Customer Deposit* (this Item must be type *Other Charge*, and default into the liability account *Unearned income*).
8. In the **Amount** field, type the *amount* of the prepayment. Let's say the amount is $2913.29 or 30% of the Estimate's total according to the company's policy.
9. Next, at the lower left section of the window, select either the **Group with other undeposited funds** or **Deposit To** option.
    a. The **Group with other undeposited funds** option uses the Undeposited Funds account. That's an account in the chart-of-accounts, that QuickBooks® creates for you. When you select the *Group with other undeposited funds* option, it uses this account to hold the funds in an undeposited status until you decide to complete the depositing of the funds in the system by placing them into a final account such as a Bank or checking account. The completing of the depositing is a separate step.
    When you use the undeposited funds account, it allows you to deposit multiple checks, instead of one at a time. Also, as we will discuss later how the undeposited funds account is used through the *Receive Payments* function, where

you can record payments from customers for completed work. These payments maybe through a credit card, which means you don't receive the moneys immediately but two or so, days later. So, it makes sense to leave the credit card type payments in the *undeposited funds* (in undeposited status) account instead of depositing them directly into the *Checking account*.

Therefore, using the undeposited funds account is a proper way to use QuickBooks®. If you elect to use the undeposited funds account by selecting the **Group with other undeposited funds** option, then you must follow with the next step and finalize the depositing of the funds. The next step is accomplished through the **Make Deposits** function, which we'll examine in the next section.

b. If you click at the **Deposit To** option, the second option, the funds go directly into a particular account you select, such as *checking* account or *savings* account. This selection allows you to deposit one check at a time.

The same options, **Group with other undeposited funds** and **Deposit To,** are available through the *Receive Payments* function, which we'll discuss later.

10. Click at the **OK** or the **Next** button to complete recording the customer *prepayment*.

After you have completed recording the prepayment, you have accomplished the following accounting:
a. The *undeposited funds* account (or the Checking account depending on your choice) has received a Debit type entry,
b. Via the use of the Item that was used in the Enter Cash Sales function, the *Unearned Income* account has received a Credit type entry.

	DR	CR
Undeposited funds account	$2913.29	
Unearned Income account		2913.29

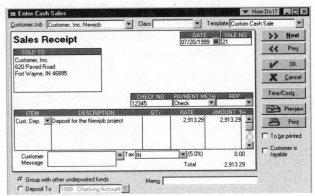

Figure 3.4 shows the customer **prepayment** (deposit).

After you record a customer prepayment into the *Unearned Income* account, the funds must remain there until your company has *earned* the income by providing the service or the product. When the income is earned, the amount in the *Unearned Income* account (the liability) can be reversed at the time you create the invoice via the *Create Invoice* function, when you are ready to bill the customer for the product or services you have provided.

Make Deposits

𝒫urpose: To complete the depositing of funds in the system that you've received from customers and others (such as refunds from vendors).

As it was described in the previous section, when you select the **Group with other undeposited funds** option in the **Enter Cash Sales** function or through the **Receive Payments** function, it automatically enters the recorded funds into the *Undeposited funds* account in the form of a *debit* type entry. These funds will stay in this account until you follow through with the depositing of the funds in QuickBooks. This step allows you to move the funds you've received from the *Undeposited Funds* account into the Bank (Checking) account. This step is accomplished via the **Make Deposit** function.

To complete the **depositing** of funds, select the following:

1. The **Activities** at the menu bar (in QuickBooks® 2000, select **Banking**).
2. Next, select the **Make Deposits** function from the drop down menu.
3. At the **Payments to Deposit** screen as shown in figure 3.5, *click* under the √ column to select the payments you want to deposit by placing a checkmark (√).
4. Next, click at the **OK** button.

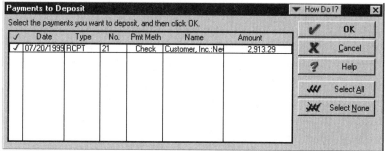

Figure 3.5 shows the payments received at the **Payments to Deposit** screen.

5. At the next screen, select the *account* you want the fund (s) to be finally deposited into, such as the Bank, Savings or another Current Asset account, at the **Deposit To** field.
6. Next, click at the **Print** button (to the right) to print either, the *deposit slip and summary* or just the *deposit summary* report, and click at the **OK** button to proceed with the printing.
7. Click the **OK** button to complete the depositing process.

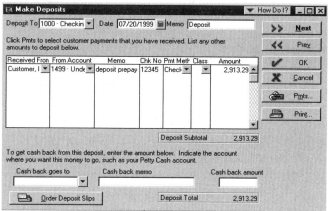

Figure 3.6 shows the **Make Deposits** screen (depositing of funds).

The accounting that has occurred after the deposit of funds is completed, is the following:

	DR	CR
Bank account	$2913.29	
Undeposited funds account		2913.29

Custom Reports

QuickBooks® provides the capability to allow you to create a *custom* report that would fit a particular purpose. For example, in the previous section we spoke about how to record properly customer prepayments. After, you record customer prepayments, you may want to keep track of all these funds so that you know the customer name, date received, check number and etc. for each amount. To create a report that would provide all the particular information about a specific need, you need to customize a new report.

For example, lets do a custom report to track customer prepayments.

To create a *Custom* report, select the following:
1. The **Reports** at the menu bar.
2. The **Custom Report** option (in QuickBooks® 2000, select **Custom Summary Report**).
3. At the new screen set the desired date either at the **Reports Dates** field or the **From** and **To** fields at the top of the screen.
4. At the **Columns** field select an option such as Month, Quarter, etc.
5. At the **Row Axis** field below, select the *Customer* option and click at the **OK** button
6. On the next screen, select the **Filters** button from the top menu bar
7. Select the *Account* option under the **Filter** column and under the **Account** field to the right click and select a particular account. In this case the *Unearned Income* account.
8. Click at the **OK** button
9. Next, click at the **Header/Footer** button from the top menu bar, and type a name for the report next to the **Report Title** filed. Click the **OK** button

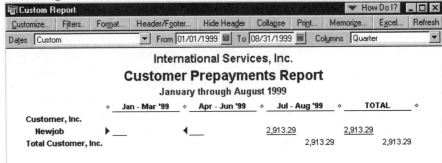

Figure 3.7 shows the custom report that tracks customer prepayments or deposits.

Keeping Cash Back

During the depositing of company funds through the **Make Deposits** function, you have the option to keep back an amount of money for such purposes as replenishing the Petty Cash fund. To keep back an amount (instead of depositing everything at the bank), you must be depositing moneys that have been received in the form of cash.

Let's assume for this example, that you have received an amount of $59.82 in cash and that you want to keep back $30, as shown in figure 3.8. The balance that is left ($29.82), you want to deposit it along with other amounts (Checks, Credit Cards, and etc.) at the bank. To record the cash you want to keep back, follow the steps below:
1. Select the **Make Deposit** function. Select the amount you want to deposit, click OK.
2. At the **Cash back goes to** field, at the next screen, select the proper account i.e. the *Petty Cash* account from the chart-of-account, as shown in figure 3.8.
3. At the **Cash back memo** field type a memo about the purpose of the cash you'll keep back.
4. At the **Cash back amount** field type the *amount* you want to keep.
5. Click at the **OK** button to complete the transaction.

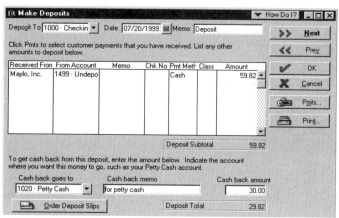

Figure 3.8 shows the cash back transaction.

The accounting that has occurred in the background is the following:

	DR	CR
Bank account	$29.82	
Petty cash account	$30.00	
Undeposited funds account		59.82

Deposit Report

After you complete the depositing of the funds, you may print the deposit report.

To print the **deposit** report, select:
1. The **Reports** at the menu bar.
2. The **Other Reports** option at the drop down menu (in QuickBooks® 2000, select **Banking**).
3. The **Deposit Detail** option for the detail report. You may view it on the screen or print it on paper by clicking at the **Print** button at the top of the screen.

Create Invoices

*P*urpose: To create customer **invoices** to bill for the sale of product or services.

Invoicing customers should be done only after the company has earned income.

Business tip: Remember that income is earned only when the company has provided the product or the service to the customer.

To **invoice** a customer for product or services that your company has provided select the following:
1. The **Activities** at the menu bar (in QuickBooks® 2000, select **Customers**).
2. Next, select the **Create Invoice** function from the drop down menu.
   Job Costing Income
3. At the *Create Invoice* screen, in order to track the **income** earned into a Job report for job costing, at the **Customer:Job** field select, by clicking, a name from the customer list. If you do not need to track the income into a job report, simply click at the customer name.
   Departmentalized Accounting
4. Next, at the **Class** field select, a Class from the Class list in order to do departmentalized accounting and track the income into a report by Class.
5. Set the following misc. Information:
   a. The *date* at the **Date** field.
   b. The Invoice *number* at the **Inv. No** field may be edited in case you need to change it otherwise, QuickBooks® automatically numbers invoices.
   c. The terms of the transaction in the Terms field. The terms determine the discount for early payment. For example, 2% 10, Net 30 indicates that the customer may pay you 2% less of the invoice amount, if payment is made within ten days from the date on the invoice.
6. Under the **Item** column, select the Item (s) from the Items list by clicking.

7. When you have finished completing the invoice, click at the **OK** or the **Next** button to proceed creating another invoice.

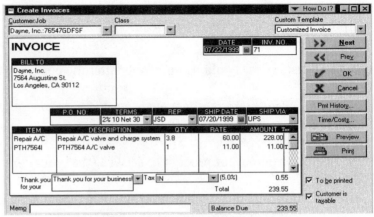

Figure 3.9 shows a completed invoice.

The accounting completed in the background by creating this example invoice, is the following:

	DR	CR
A/R account	$239.55	
Sales – Parts		11.00
Sales - Services		228.00
Sales tax payable		.55

Remember any time you create an invoice that involves an Item that is **Inventory Part** type; there is an additional transaction that occurs in the background that you don't see. In addition to the transaction listed above, there is also the following transaction that takes place at the time of the sale of an inventory item (which is driven by an Inventory Part type Item), during the creation of the invoice. The system records the reduction of the inventory asset account and the "cost" of the sold inventory item.

Here is the accounting (assuming, for this example the cost of one unit sold is $6.5 each):

	DR	CR
COGS account	$6.5	
Inventory account		6.5

125

The Cost of Goods Sold account (COGS) though it is COGS type (there is a type called COGS in the chart-of-accounts), it is still an expense "type" of account and its deducted from the income on the P/L statement.

Note: The *Sales - Parts* account (income account) has been affected by the Inventory Part Item.
The *Sales - Service* account (income account) has been affected by the Service Item.
The Sales Tax type Item has affected the *Sales Tax Payable* account (liability).

The A/R Report

After you create invoices, there is available to you the Accounts Receivable (A/R) report. On a weekly basis, you should print the **A/R** report in either *detail* or *summary* format. Its purpose is to inform you of all the amounts owed to your company from the various customers.

Business tip: The A/R report plays an important role in the cash flow of your business and it should be printed on a weekly basis. This report shows the amounts that are due from customers, and the age of each amount.

To print the **A/R** reports select:
1. The **Reports** at the menu bar
2. The **A/R Reports** at the drop down menu (in QuickBooks® 2000, select **Customer & Receivables**).
3. And the **A/R Aging Detail** or the **Summary** report

Converting Estimates to Invoices

If you have created an *estimate* already for a customer and proceed to invoice (the same customer) for the product or the services you have provided, QuickBooks® automatically provides you with the reminder as shown in figure 3.10: This Customer:Job has an estimate. Do you want to create this invoice based on an estimate?

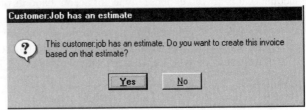

Figure 3.10 shows the invoice reminder.

After you have received this reminder you have two options: the **Yes** or **No** option:
1. The **No** option, allow you to create another Invoice, for the same customer, not related to the estimate.
2. The **Yes** option instantly converts the Estimate into an Invoice.

Once you are in the invoice function, you may change *quantities* and *rates* as you wish.
Note: If you have received a *prepayment* from the customer you are invoicing, it is in this stage that you need to *apply* the prepayment against the invoice.
When you apply the prepayment against the invoice amount, you are reducing the amount the customer will owe you as a result of the billing.

To *apply* the customer prepayment, follow these steps:
1. Go to the button of the Invoice (under the very last Item) and click with the cursor under the **Item** column.
2. Select the *customer prepayment* Item we spoke on chapter 1, from the Items list
3. In the **Amount** column, type the *amount* of the prepayment (the example we had above when we discussed the 30% prepayment) in this case $2913.29, with the negative sign in front (use the hyphen). Next, click anywhere with the left mouse button. And instantly, you have accomplish three things:
   a. By typing the amount of the prepayment in the invoice function with a negative number, it instantly reverses the amount from the *liability* account (Unearned Income)
   b. Next, the *Accounts Receivable* account will receive the adjusted amount of the invoice. Which is: The amount of the Invoice – the Prepayment amount = amount due.
   c. Next, through the use of the Items listed on the Invoice, you will record the earning of the *income* that the company has earned (via the default account in each Item), and thus when you post the invoice you instantly update the P&L

report.
4. Next, click at the **OK** button to post the Invoice.
In figure 3.11, please notice the negative amount of the prepayment at the end of the invoice.

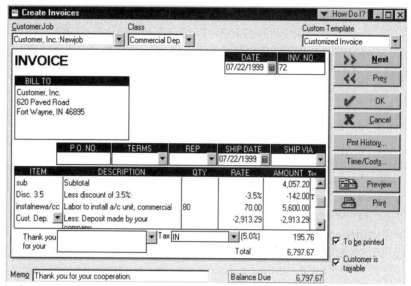

Figure 3.11 shows an invoice with the customer prepayment *applied*.

Progress Billing

*P*urpose: To invoice in incremental amounts that are based on a percentage or $ amounts, instead of the full amount of the total Estimate (or contract).

Progress billing is based on the progression of work that has been completed and is mostly common in the construction industry where a project can be divided into various phases. As a phase is completed, the customer can be invoiced for that portion of the work only.

To be able to do **Progress** billing, you first must complete the following:
A. Set the *Preferences* to accommodate progress billing

B. Have an *Estimate* already created

To set the *Preferences*, select the following:
1. The **File** at the menu bar (in QuickBooks® 2000, select **Edit**).
2. Next, select the **Preferences** option
3. In the Preferences screen, select the **Job & Estimate** Preference.
4. In the **Company Preference** tab, click at the **Yes** option under the **Do You Do Progress Invoicing?**
5. Next, click at the **OK** button.

To invoice a customer using **Progress Billing**, follow the same steps that are outlined at the beginning of the *Create Invoices* section. The only difference is that after you select a customer or job name at the **Customer:Job** field in the invoice QuickBooks® provides you with an additional screen that allows you to select from *three* options as shown in figure 3.12:
1. *Create invoice for the entire estimate (100%).* This option creates an invoice for the *entire* estimate.
2. *Create invoice for a percentage of the entire estimate.* This option takes each Item and multiplies it automatically by the percentage you type in the **% of estimate** field.
3. *Create invoice for selected items or for a different percentage of each item.* This option allows you to type different *percentages* or *amounts* for each Item on the invoice.

Select one of the options by clicking. Next, click at the **OK** button to create the invoice that is based on progress billing and according to the option you chose above.
Once you are in the invoice function, you can change any amount you may choose.

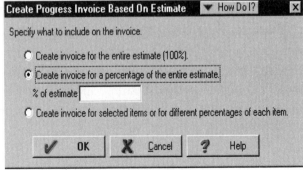

Figure 3.12 shows the **Create Progress Invoice** screen.

129

Receive Payments

𝒫urpose: To record payments received from customers as a result of invoicing.

When you receive a payment from a customer that is due to an invoice, the amount must be recorded via the **Receive Payments** function.

To record a customer **payment**, select the following:
1. The **Activities** at the menu bar (in QuickBooks® 2000, select **Customers**).
2. The **Receive Payments** option at the drop down menu
3. And at the **Customer:Job** field select a customer or job *name* from the customer list. Once the customer is selected, the screen shows the following:
   a. The *balance* that is outstanding at the **Balance** field.
   b. The open *invoice* (s) at the bottom of the screen as shown on figure 3.13.
4. Next, type the *amount* of the check in the **Amount** field in the upper right corner.
5. Next, select the *Payment Method* at the **Pmt. Method** field by either clicking on the arrow, to select the payment method, or type **CH** for check.
6. Next, type the *check number* at the **Check No** field.
   Note: All this information is relevant and goes into the customer history.
7. After you type the amount of the payment, you must **apply** it to the Invoice.
   To apply the amount to the Invoice, it can be done by:
   a. Clicking at the **Auto Apply** button. This option applies the payment to the oldest invoice and proceeds to apply the balance to the rest of the invoices.
   b. **Manually** clicking next to the invoice (s) listed under *the Outstanding Invoices/Statements Charges* section.
   *Note: If you don't apply payments to invoices, the **unapplied** payment amounts show up in the Accounts Receivable (A/R) report as negative amounts.*
   If you forget to *apply* a payment at the time of recording it, you may go through the Receive Payments function, find the unapplied payment, and apply it to the invoice at a later time.
8. Next, at the lower left section of the window, select either the **Group with other Undeposited funds** or **Deposit To** option.
   c. The **Group with other undeposited funds** option uses the Undeposited funds account. That's an account in the chart-of-accounts, that QuickBooks® creates for you. When you select the *Group with other undeposited funds* option, it uses this account to hold the funds in an undeposited status until you decide to complete

the depositing of the funds in the system by placing them into a final account, such as a Bank or checking account.

The completing of the deposit is a separate step as we have discussed earlier at the **Make Deposits** section.

When you use the undeposited funds account, it allows you to deposit multiple checks, instead of making one-check deposits.

If a customer pays you with a credit card you don't receive the moneys immediately until two or so days later. It makes sense to use the *undeposited funds* account because these types of payments should stay in an undeposited status instead of depositing them directly into the Bank (Checking) account. Therefore, using the undeposited funds account is a proper way to use QuickBooks®. If you elect to use the undeposited funds account by selecting the **Group with other Undeposited funds** option, then you must follow with the next step and finalize the depositing of the funds. The next step is done through the **Make Deposits** function.

a. If you click at the **Deposit To** option, the funds go directly into a particular account you select such as *checking* account or *savings* account. This selection allows you to deposit one check at a time.

These options, **Group with other undeposited funds** and **Deposit To,** are also available through the *Enter Cash Sales* function that we'll be discussing later.

9. After you record the payment, select the **OK** or **the Next** button to *post* the payment.

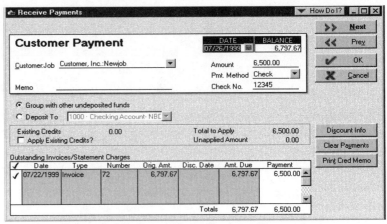

Figure 3.13 shows the customer Payment.

Let's say that you have received a payment for invoice # 72 that is shown on figure 3.11, and that the amount of the payment is for $6500.

a. If the payment is recorded via the **Receive Payment** function as shown in figure 3.13, the following accounting takes place:

	DR	CR
Undeposited funds account	$6500	
A/R account		6500

b. When you complete the *depositing* of the monies received in the system through the **Make Deposits** function as shown in figure 3.14, the following accounting takes place:

	DR	CR
Checking account	$6500	
Undeposited funds account		6500

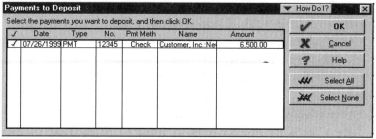

Figure 3.14 shows the deposit of $6500.

To print a **report** that shows the payments you have received, select the following:
1. The **Reports** at the menu bar
2. The **Transaction Detail Reports** and next, select the **By Date** option
   (In QuickBooks® 2000, select **Reports/Customers & Receivables** and **Transaction List by Customer**).
3. Click at the **Filters** button above
4. Select the **Transaction Type** option under the Filter heading, and to the right, at the **Transaction Type** option, click and select **Payment**.
5. Click at the **OK** button and at the next screen, adjust the dates at the **From** and **To** fields.

Write-off Bad Debt

In the event that you have invoiced a customer for providing product or services and later you receive a partial payment or perhaps no payment at all, in your system you will have an outstanding or open invoice.
Business tip: Do not delete invoices or change the amounts on outstanding invoices. To remove uncollectable amounts from your books, you must write them off.

You may write-off an open invoice only after you have exhausted all possible options of collecting your money.

For example, let's say invoice # 72 as shown in figure 3.11, had a total amount of $6797.67 and you have received a payment of $6500 as shown on figure 3.13, and the $297.67 is left outstanding (unpaid).

To write-off **bad debt**, follow these steps:
1. Select the **Activities** at the menu bar (in QuickBooks® 2000, select **Customers**).
2. Select the **Create Credit Memo/Refunds** function and click once
3. In the Create Credit Memo screen, select the *customer name* at the **Customer:Job** field
4. At the **Item** column, select the Item that would help you create the credit memo.
   Note: The Item you need must be *Service* type and make it to default into an expense type account that you could name as *Bad debt*.

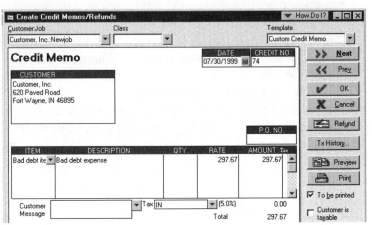

Figure 3.15 shows the Credit Memo.

5. Click at the **OK** button.
6. Next, select the **Receive Payments** function again and
7. Select the customer *name*
8. Click at the **Apply Existing Credits** field
9. Click next to the outstanding invoice to *apply* the credit. And this action will eliminate the outstanding invoice amount from the Accounts Receivable report.

The accounting that has occurred in the background is as follows:

	DR	**CR**
Bad debt account	$297.67	
A/R account		297.67

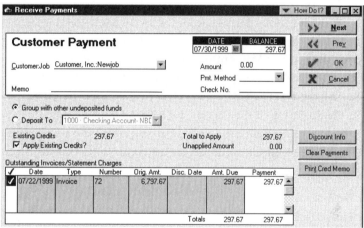

Figure 3.16 shows the applying of the *credit* to the outstanding *invoice* (unpaid).

Create Credit Memos/Refunds

𝒫urpose: Use the **Credit Memo** function to issue a *Credit* to a customer or a *Refund* check.

Remember: A credit that you issue to a customer decreases the customer *balance* and the *Accounts Receivable* account. It also decreases the *Income* account that the Item is defaulting into.

134

Issue Credits to Customers

To issue a **Credit** to a customer for returned product or faulty service, select the following:
1. The **Activities** at the menu bar (in QuickBooks® 2000, select **Customers**).
2. Next, select the **Create Credit Memo/Refunds** function from the drop down menu
3. In the Create Credit Memo screen, select the *customer name* at the **Customer:Job** field
4. At the **Item** column, select the Item that would help you create the credit memo. The Item you need may be:
    a. *Service* type that can be used in the event that faulty **services** have been provided
    b. *Inventory Part* type that can be used for returned **product**
5. Next, type the *quantity* and the *amount* in the respective fields
6. Click at the **OK** button to post the transaction.
7. Next, select the **Receive Payments** function and
8. At the *Receive Payments* function, select the customer name
9. Next, click at the **Apply Existing Credits** field
10. Place a checkmark (√) by clicking next the outstanding invoice to apply the *credit*. And that will reduce or remove an outstanding invoice from Accounts Receivable (A/R).

Issue Refunds to Customers

To issue a **Refund** to a customer for returned product, select the following:
1. The **Activities** at the menu bar (in QuickBooks® 2000, select **Customers**).
2. Next, select the **Create Credit Memo/Refunds** function
3. In the Create Credit Memo screen, select the *customer name* at the **Customer:Job** field
4. At the **Item** column, select an Item from the Items listing that would help you create the Refund.
5. Next, type the *amount* in the **Amount** field
6. Click at the **Refund** button to create a check
7. At the **Write Checks** screen, click at the **OK** button to create the check

8. Back at the *Create Credit Memos/Refunds* screen, <u>click at the **Cancel** button</u>
9. Next, select the **Receive Payments** function again
10. In the *Receive Payments* function, select the customer name
11. Next, click to place a checkmark (√) at the **Apply Existing Credits** field
12. Click to place a checkmark (√) next TO the outstanding invoice to *apply* the *refund*.

This action will reduce the customer outstanding balance and will remove the invoice from Account Receivable.

Enter Cash Sales

𝒫urpose: To record *income* your company has earned and the *payment* it has received, both in one step.

The **Enter Cash Sales** function allows you to record the **sale** of product or service (that's income that has been earned) and the **payment** at the same time in order to accommodate business transactions where after you provide the service or the product, you get paid immediately. Therefore, there is no need to create an invoice.

Various types of businesses that have the need to achieve the above objective, that is, to record the income and the payment at the same time can use the **Enter Cash Sales** function, such as:
1. Service type companies to record the sale of services.
2. Retail or other companies that may sell product.
3. Non-profit organizations to record *donations (gifts)* or *grants* they may receive.

A. **For Service Companies:**
   If your company provides services and your policy is that you receive payment upon completion of providing the services, your employees ought to return with a method of payment such as *check* or *credit card*.

   To record the sale of a <u>service</u> and the <u>payment</u>, select the following:

1. The **Activities** at the menu bar (in QuickBooks® 2000, select **Customers**).
2. The **Enter Cash Sales** function.
3. In the **Customer:Job** field select, the customer name for which you provided the service.
4. In the **Class**, select a Class to track the income into a Department.
5. Next, set the *date* in the **Date** field.
6. In the **PAYMENT METH** field, select the method of payment such as Master Card, check, etc.
7. In the **Item** column, select a **Service** type Item from the Item listing that would allow you to record the service your company has provided. The Item will appear on the screen with a default rate i.e. $50 per hour.
8. In the **QTY** field, type a quantity, i.e. 5 hours (of service). Next click anywhere with the mouse and QuickBooks will multiply the quantity with the rate, 5 X 50 = $250. So by completing this transaction, we're accomplishing two things: the income earned via the use of the Item and the method of payment.
9. Next, allow QuickBooks® to default to the **Group with other Undeposited funds** field we have explained in the *Make Deposit* section.
10. Next, click at the **OK** or the Next button to post the transaction

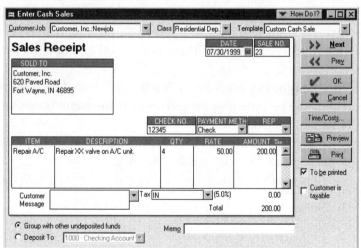

Figure 3.17 shows the **service** that has been provided to a customer and the payment received.

In one step, we've recorded the *income* (via the account that the Item is defaulting into) and the *payment* we've received. Another option that's available to you to accomplish the

same task, and the same accounting, requires two steps:
1. You first create an *invoice*
2. Second, you apply the *payment* to the invoice.

Enter Cash Sales function is capable of printing a **sales receipt**. In the event the customer requests a receipt from you, click at the **To Be Printed** button to the right of the form, and you can print a sales receipt.

B. **For Product Companies:**
If you company is selling a **product**, such as retail business, to record the sale of the product and the payment, follow all the steps explained in the Service example above, except in step #7, select an *Inventory Part* type Item to record the sale. Using this function, you can accommodate walk-in customers by recording each sale as it occurs.

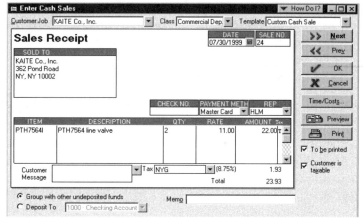

Figure 3.18 shows the sale of **product** and the payment.

If you are using a POS program to manage the sales, the inventory, and Accounts Receivable part of your business, you can use QuickBooks® for all the other tasks, such as the Payroll, A/P and Financial reporting. In this case, you may record the entire day's or the week's sales and payment activity in a summary format.

Using the **Enter Cash Sales** function, you may, for example, record an entire day's (or week's) sales by using **Service** type Items (because you are recording summary activity, the Items must be Service type). The sales can be recorded for each product you sell or method of payment. To record by product, create an Item for each product (or Service). To record by method of payment, simply select the method at the **Payment Meth** field.

The Items you create must be defaulting into Income type accounts. Then using these Items, you may record the sales and payments for an entire day or week.

C. **Non-profit Organization:**

If your organization is non-profit and you need to record monies that come-in in the form of *gifts*, *donations* and *grants*, follow **all** the steps outlined in example **A** above.

For every type of "gift", you must create an Item that is **Service** type. Each of these Items must default into an **Income** type account as shown in figure 3.19 (please refer to the Items section, in Chapter 1 where we explain how to create *Service* type Items).

In the event however, that you receive a "gift" that has been designated by the Giver to be used only for a very particular purpose such as to build a new building, etc., you may want to record such a fund into a new Equity type account and not into an Income type account (please consult with your Accountant before you record such a fund).

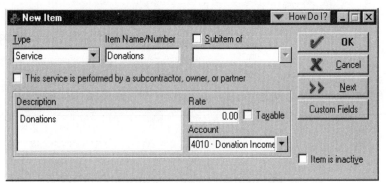

Figure 3.19 shows a **Service** type Item.

After you record the receiving of funds, remember to follow through and complete the depositing of the funds in the system via the **Make Deposit** function.

Enter Statement Charges

*P*urpose: Use the **Customer** and **Accounts Receivable** register to record transactions that pertain to customers.

Through The **Enter Statement Charges** function, you can *enter* customer *charges* that you want to appear on the next billing statement. You can also *edit* or change prior statement charges.

The **Enter Statement Charges** function shows all the Accounts Receivable transactions pertaining to a particular customer.

To enter customer **charges**, select the following:
1. The **Activities** at the menu bar (in QuickBooks® 2000, select **Customers**).
2. The **Enter Statement Charges** function at the drop down menu, and click once
3. Select the customer name under the **Customer:Job** column.
4. Type a number or reference at the **Number** field
5. Select an **Item** under the Item column
6. Type a *quantity* under the **Qty** field
7. Click at the **Record** button to complete and to post the transaction.

Figure 3.20 shows a **statement charge** for a customer.

140

Assess Finance Charges

*P*urpose: To assess (or charge) **finance charges** to customers with an overdue balance.

The **Assess Finance Charges** function lists all the customers who have an outstanding balance with your company and it allows you to assess finance charges that are based upon the outstanding amounts due.

Assessing finance charges should be done at the *end* of each month. QuickBooks® has the capability of printing invoices for the finance charges you access. After you assess finance charges, you should send *monthly statements* to your customers.

To assess **finance** charges, select the following:
1. The **Activities** at the menu bar (in QuickBooks® 2000, select **Customers**).
2. The **Assess Finance Charges** option from the drop down menu
3. At the **Assess Finance Charges** screen, and if you ARE entering this screen for the first time, select the **Settings** button by clicking once. Now, you are in the Preferences.
4. In the **Preferences** screen, set the following fields:
   a. In the **Annual interest rate**, type the annual rate you want to charge overdue customers
   b. In the **Minimum finance charge**, type a min. charge that your policy may allow
   c. At the **Grace period** field, type a number of days (or zero) before finance charges apply
   d. At the **Finance charge** account field select an account from the chart-of-accounts. This account must be **Other Income** type.
   e. Assess finance charges on overdue finance charges.
   f. At the **Calculate finance charges from** field select according to your company policy
   g. Click at the **OK** button to save the settings.
5. Back at the **Assess Finance Charges** screen, **click** or select the **Mark All** button to place a checkmark (√) next to each customer's balance to *assess* finance charges.
6. After you finish, click at the **OK** button.

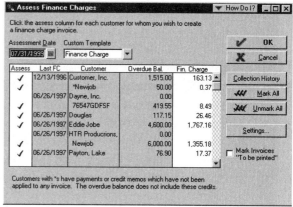

Figure 3.21 shows the **Assess Finance Charges** screen.

After you **assess** finance charges, the following activity takes place:
  a. The customer's outstanding balance *increases*
  b. The **Accounts Receivable** account (A/R) *increases*
  c. Your company's income *increases*
   The following accounting takes place in the background:

	DR	CR
A/R account	$X	
Other Income account		X

Where X = the transaction amount.

Note: Other Income is simply the name of the account that allows you to record the income from the finance charges. This account must be an *Other Income* type account.

Create Statements

*P*urpose: To create monthly statements that you can send to customers that show the outstanding balances owed to your company.

After you assess finance charges, you can create and print the customer statements.

Statements can be created for just one or for multiple customers.

To create **statements**, select the following:
1. The **Activities** at the menu bar (in QuickBooks® 2000, select **Customers**).
2. The **Create Statements** from the drop down menu.
3. At the **Select Statements to Print** screen, set the following:
   a. The dates at the **Dates From** and **To** fields, to indicate the date range you want the statement to cover. For example, if you want to print just July's activity, select 7/1/99 in the **Dates From** field and 7/31/99 in the **To** field.
   b. The **Statement Date** field appears at the top of the statement. This date determines the due date of the charges.
   c. The **For Customers** field, allows you to select whether you want *all* customers or just *one* customer to receive a statement.
4. Next, select by clicking the **Preview** button to preview the statements before you print them
5. After previewing, you may print by clicking at the **Print** button.

Customer Bad Checks
(Insufficient funds checks)

Often in the course of doing business, there are times when you may have to face a situation where after you have done your best in providing a service or a product to a customer, you receive a *disappointment* instead of a *payment*. Here is a possible scenario.

After you provide a service (or a product) to one of your customers, you create the **invoice** to bill this customer. The customer then follows through by sending you a **payment**. When you receive the payment check, you record it via the **Receive Payments** function. Next, you deposit the payment check at the Bank. A few days later, you receive a note from the Bank indicating that the check you have deposited is a bad check or it's designated as "insufficient funds".

After you've recorded the payment, here is what happened to the accounts in your system that have been affected by this payment:
1. Your **Checking** account has been *increased*.
2. Your **Accounts Receivable** (A/R) has been *decreased* because you have applied the payment to the Invoice.
3. The Bank has charged your account a *fee* because they've tried to collect for you.
4. The *Invoice* that you have applied the payment has been moved into the history.

To correct the situation described above, you need to do the following:
1. *Decrease* the **Checking** account.
2. *Increase* the **Accounts Receivable**.
3. Record the *expense* of the Bank fee and perhaps, *bill* the customer back for the fee that the Bank has charged you.

Note: Please remember to not change the amounts on the old invoice.

After you receive the notice from the bank that a certain check is designated as insufficient funds, proceed with the following steps:
1. Select the **Activities** at the menu bar (in QuickBooks® 2000, select **Customers**).
2. Next, select the **Create Invoice** function from the drop down menu
3. At the **Customer:Job** field select, the customer name.
4. At the **Item** column, select the *Items* that would help you manage the bad check situation. To manage bad checks, you need two Items:
    a. Item # 1. This Item must an **Other Charge** type item and be defaulting into the **Checking** account. This Item will decrease the Checking account (it has to go down, because when you record a payment, the Checking account went up).

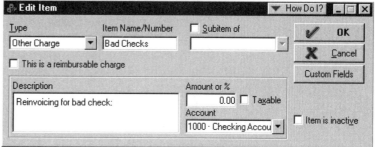

Figure 3.22 shows Item # 1.

b. Item # 2. This Item must be **Other Charge** type also and be defaulting into an account that is **Other Income** type with a possible name such as *Misc. Other Income*. By using this Item on the Invoice, you will be able to charge back the customer for the Bank fee and record it as *income* in your books.

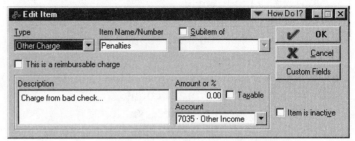

Figure 3.23 shows Item # 2.

5. At the **Amount** field:
   a. Type the *amount* of the bad check, let's say it was $700
   b. Type the *amount* of the Bank fee that you may want to charge the customer back. Let's say the bank fee was $25.
6. Click at the **OK** button to post and print the invoice.

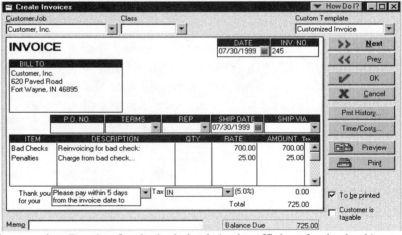

Figure 3.24 shows the complete **Invoice** for the bad check (or insufficient funds check).

7. Next, through either the **Make Journal Entry** or the **Reconcile** function at the time of reconciling the Bank account statement, record the Bank fee.

145

If you record the Bank fee via the **Make Journal Entry** function, this is how you must enter it:

	DR	CR
Bank charges account	$25	
Checking account		25

You have finished correcting the data in your system after the bad check.

Chapter 4

Payroll

The purpose of this chapter is to help you prepare the Payroll Items, keep track of time, do the Payroll, print paychecks, and pay the payroll taxes. Also, how to prepare the 941 and 940 payroll tax forms as well as the W-2 forms.

In this chapter we'll examine the following QuickBooks® functions:

- **Create Payroll Items**
- **Enter Employees**
- **Track Time**
- **Pay Employees**
- **Pay the Liabilities and Taxes**
- **Payroll Forms**

Payroll Items

*P*urpose: Is to create the **Payroll Items** that are needed in order to do payroll according to the company's needs and meet Federal and State government regulations.

The **Payroll Items** are used in the payroll function to create paychecks and pay the company employees. Through them, you can accomplish the following tasks:
   a. *Pay* the salaried and hourly employees
   b. *Deduct* from the paychecks for child support, garnishments, union dues, insurance, etc.
   c. *Withhold* taxes for the Federal, State and Local governments
   d. Define the *accounts* that needed to be used in the payroll and thus do proper accounting
   e. Define the *names* and the *addresses* that need to be printed on the checks and paychecks for employees and the various agencies that you need to send checks to pay the taxes.

Payroll Items use list

The following list, you will learn about:
   a. The possible *purpose* of the Payroll Item you may need to create
   b. The *type* you may need to select from the Payroll Items list
   c. The *account* (from the chart-of-accounts) you may need to select for the Item to default (the account you select and assign to the Item, receives the amount of expenses, liability, or assets that pertain to payroll).

Table 2

If you want:	Use Item Type:	Add/Ded. to Gross/Net:	Account to use:
Bonus Pay	Addition	Gross	Expense
Mileage Reimbursement	Addition	Net	Expense
When Company Issues Employee Advances	Addition	Net	Asset
When Employees Pay			

Back the Advance	Deduction	Net	Asset
Tips	Addition	Gross	Expense
EIC	Addition	Net	Expense
Union Dues	Deduction	Net	Liability
Cafeteria Plans Deferred Comp.	Deduction	Gross	Liability
Garnishments	Deduction	Net	Liability
Employee Donations (to charities)	Deduction	Net	Liability
Health/Life Insurance (paid by Employees)	Deduction	Net	Expense
Health/Life Insurance (paid by Company)	Company Contrib.	N/A	Expense
Health/Life Insurance (Coop)	Deduction	Gr./Net (depending on plan)	Expense
State or Local Tax W/hh.	Deduction	Gross	Liability
State or Local Tax paid by Company	Company Contrib.	N/A	Expense/Liability

To create new **Payroll Items**, select the following:

1. The **Lists** at the menu bar
2. The **Payroll Items** option from the drop down menu.
3. Next, select the **Payroll Item** button at the *Payroll Item List* screen
4. Next, select the **New** option.
5. In the **Add new payroll item** screen there are the payroll item **type** selections. Select the type you want and click at the **Next** button.
   a. The very first type of Payroll Item is the **Wage**. This is the type you need to have in order to pay your employees using QuickBooks®. It will allow you to create paychecks for your Salaried and Hourly employees.

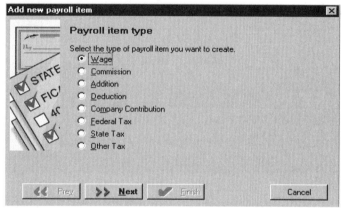

Figure 4.1 shows the **Add new payroll item** screen.

- At the **Wages** screen, select one of the following: **Salary, Hourly, Sick Pay** or **Vacation Pay** payroll Item and click at the **Next** button to continue.
  The **Salary Wages** type payroll item is designed to allow you to pay salaried type employees, such as the Officer of the company or an office employee. You need to create one or maybe two of this type of Items for employees that work in the office.
  The **Hourly Wages** type payroll item is designed to allow you to create a payroll item to pay the hourly employees. You need to create one or maybe two of this type of Items for employees that work outside of the office, such as field or shop employees.
  The **Vacation Pay** type, allows you to have an Item to pay employees when they take vacation. You need to have only one of this type.
  The **Sick Pay** type, allows you to have an Item that you use to pay employees when they are out due to sickness. You need to have only one of this type.

- After selecting the Wage type, click at the **Next** button and in the next screen, type a *name*, for this new Item and click at the **Next** button. The name you type is the ID of the item. All items must have a name. Click at the **Next** button to continue.
- In the next screen, select an *account* that is Expense type. This account will be used by the Item to record the expense of the salary wages (in this case) when you'll use Item to pay an office employee. The proper account to choose as the default account is the **Office Wages** account.
- Next, click at the **Finish** button, to complete the creation of the item.

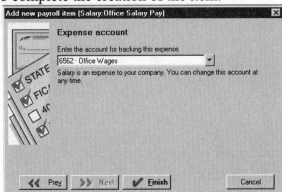

Figure 4.2 shows the **Salary Wage** type payroll Item.

To create **Hourly Wages**, **Vacation Pay** or **Sick Pay** type Items, follow the same steps used to create the **Salary Wage** type except designate a different expense account for each Item.

b. The next Item below is the **Commissions** type. This is the type that you need to create in order to pay commissions to employees such as sale people. Click at the **Next** button to continue:
- Next type a *name* for the new Item such as Commissions, and click at the **Next** button to continue.
- In the **Expense Account** field, select an account, such as *Commission Expense*. It's the account that will receive the expense of the commission. Click at the **Next** button.
- Next, type the default *rate* that you will be paying commission, i.e. 12%. The rate can be adjusted via the Item or during the payroll process (when you do payroll). This simply is the rate that will come up any time you use this item.

- Click at the **Finish** button to complete the creation of the item.

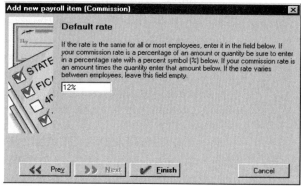

Figure 4.3 shows the **Commission** Item.

c. Next is the **Addition** type Payroll Item, the third option on figure 4.1. It is the type that would allow you to add an *additional pay* to salary and wages. Additional pay could be bonuses, mileage reimbursement, or other miscellaneous reimbursements.

To create an Addition type Item, follow steps 1-4. At the 5th step, select the **Addition** type and click at the **Next** button to proceed to the next step.

Example for **Mileage Reimbursement**, an Addition type, Item:
- Type a *name* such as Mileage Reimbursement, and click at the **Next** button.
- In the next window, select an *account* that is expense type; in this case the *Auto Expense* is the proper account. Click at the Next button.
- The **Tax tracking type** screen is where you determine whether to track this Item for tax purposes.
Note: This is a tax issue and if your are not sure whether a certain pay should be tracked for tax purposes, please consult with your accountant.
Mileage reimbursement most likely should not be tracked. The proper selection is the option **None**. Continue by selecting the **Next** button.
- In the next screen, make sure that the payroll Items under the Payroll Item field do <u>not</u> have a checkmark next to them. Click at the **Next** button.
- At the **Calculate based on the quantity** screen, click at the **Based on Quantity** field. That will allow you to type a quantity each time you do the

payroll. The quantity will be the amount of miles each employee has driven. And once QuickBooks® sees a mileage quantity, then it multiplies the *quantity* and the *rate* and establishes the pay for mileage.
- At the **Gross vs. net** field, select the **net pay**, so that the mileage pay would not be taxed. Continue by selecting the **Next** button.
- At the **Default rate and limit** screen, type the *rate* that your company reimburses employees for mileage.
  Please, consult with your accountant for the rate that is allowed by the Internal Revenue Service.
  At the **Annual Limit** field, you can set an annual limit that will cause QuickBooks® to stop this Addition from occurring forever. This field may be applicable in other cases but it doesn't apply to mileage reimbursement.
  Click at the **Finish** button to complete this Item.
  You need to create only one of this type of Item.

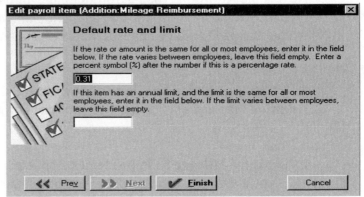

Figure 4.4 shows the **Mileage Reimbursement** Item.

Example for **Bonus,** an Addition type, Item:
- Type a *name* such as Bonus, and click at the **Next** button.
- In the next window, select an *account* that is expense type in this case the *Bonus Expense* is the proper account. Click at the **Next** button to continue.
- The **Tax tracking type** screen is where you determine whether to track this Item for tax purposes.
  Note: This is a tax issue, if your are not sure whether certain pay should be tracked, please consult with your accountant.
  Bonus pay, most likely should be tracked because it is taxable. The proper

selection therefore, should be the **Compensation** option. Click the **Next** button to continue.
- In the next screen, make sure all these payroll Items under the **Payroll Item** field <u>do</u> have a checkmark next to them as shown in figure 4.5. Click at the **Next** button.

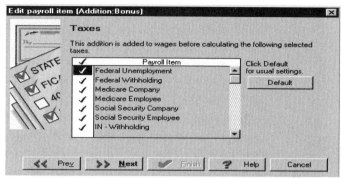

Figure 4.5 shows the **Payroll Item** screen.

- At the **Calculate based on the quantity** screen, do not click at the **Based on Quantity** field.
- At the **Default rate and limit** screen, leave the *rate* and the **Annual Limit** fields blank so that this Item can be used for all your employees by simply changing the amount of the bonus for each employee during the payroll. Next, click at the **Finish** button to complete this Item.
You only need to create one of this type of Item.

d. The next Item type is **Deduction**. This type allows you to deduct from employee paychecks for such purposes as Child support, Garnishment, 401 (k) plans, Insurance health plans, etc.

To create a **Deduction** type Item, follow steps 1-4 outlined at the begin of this section. At step # 5, select the **Deduction** type and click at the **Next** button to proceed to next step.

Example for a **401(k),** Deduction type, Item:
- Type a *name* such as 401(k), and click at the **Next** button.
- At the next screen, type the *name* of the company that manages your 401(k)

154

program. This name will print on the check that you will prepare for the company that manages the 401(k) plan.

In the space below, select an *account* that is *liability* type and in this case the *401k Payable* account is the proper one. This is the account that will accumulate and retain the deduction, until you print a check for the finance company.

Click at the **Next** button to continue.

- The **Tax tracking type** screen is where you determine whether to track the Item for tax purposes.

  Note: This is a tax issue and if your are not sure whether a certain deduction should be tracked, please consult with your accountant.

  A deduction for 401(k) plan most likely should be tracked. The proper selection is the **401(k),** option. Continue by selecting the **Next** button.

- Based on the selection of the previous step, this new screen comes with default settings. Make sure you set the Payroll Items according to government regulations.

Note: When an Item has a checkmark (√) next to it, the deduction will be deducted <u>first</u> and then the tax will calculate. An Item without a checkmark will calculate the tax <u>first</u> and then the deduction will be deducted.

Please consult with your accountant for the proper position of the Payroll Items. Also, after you set the Payroll Item position in this screen, run a sample payroll for an employee and check the figures manually to insure accuracy.

A possible position of the payroll Items <u>maybe</u> as shown in figure 4.6.

Click at the **Next** button to continue.

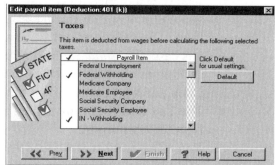

Figure 4.6 shows a possible setting of the Payroll Items for the 401(k) deduction Item.

- At the **Calculate based on the quantity** screen, do not click at the **Based on Quantity** field. Click at the **Next** button to continue.
- At the **Default rate and limit** screen, type the *rate* that you want to deduct for the 401(k) plan, i.e. 5%.
  At the **Annual Limit** field, you can type an *annual* limit that will cause QuickBooks® to stop this Deduction from occurring. This limit is according to the government regulations.
  Please, consult with your accountant for the **correct** *annual limit.*
  Click at the **Finish** button to complete this Item. You only need one of this type of Item.

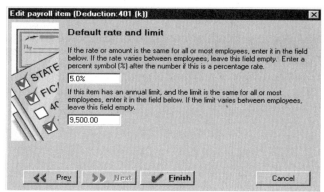

Figure 4.7 shows the 401k **Default rate and limit** screen.

Example for **Child Support,** a Deduction type Item**:**
- Type a *name* such as **Child Support** and click at the **Next** button.

156

- At the **Enter name of agency** field, type the *name* of the agency you need to send the payments for the child support. This name will print on the check that you will prepare for the agency.
  In the **Liability account** field, select an *account* that is *liability* type and in this case, the *Child Support Payable* account. This is the account that will accumulate and retain the deduction, until you print a check for the agency. Click the **Next** button to continue.
- The **Tax tracking type** screen is where you determine whether to track this Item for tax purposes.
  Note: This is a tax issue and if your are not sure whether a certain deduction should be tracked, please consult with your accountant.
  A deduction for child support most likely should <u>not</u> be tracked. The proper selection is the option **None**. Continue by selecting the **Next** button.
- Based on the selection of the previous screen, this new screen comes with default settings. Please make sure that you set the Payroll Items according to government regulations.
  Note: When an Item has a checkmark (√) next to it, the deduction will be deducted <u>first</u> and then the tax will calculate. An Item without a checkmark will calculate the tax <u>first</u> and then the deduction will take place.
  Note: Please, consult with your accountant for the proper position of the Payroll Items. Also, after you set the Payroll Item position in this screen, run a sample payroll for an employee and check the figures manually before you finalize your payroll.
  A possible position for child support may be as shown in figure 4.8.
  Click at the **Next** button to continue.

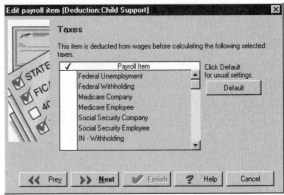

Figure 4.8 shows a possible setting for the **Child Support** deduction Payroll Item.

157

- At the **Calculate based on the quantity** screen, do not click at the **Based on Quantity** field. Click at the **Next** button.
- At the **Gross vs. net** field select the **net pay** option and click at the **Next** button to continue.
- At the **Default rate and limit** screen, type the *rate* or *amount* that you want to deduct for child support and leave the **Annual Limit** field blank.
  Click at the **Finish** button to complete this Item.
  You may need to create an Item of this type for every employee.

Example for a **Garnishment**, a Deduction type, Item:
  To create an Item to deduct for a Garnishment, follow all the steps outlined in the Child Support deduction Item except you more likely have to set an **annual** limit.

Example for **Group Health** plan, a Deduction type Item:
- Type a *name* such as **Group Health Insurance** and click at the **Next** button.
- At the **Enter name of agency** field, do not type a *name*. If you leave this field blank, QuickBooks® will not prepare a check for the amounts deducted, and you do not need to prepare a check if you have a coop program. As you will see in the next field, you may direct the amounts deducted from the payroll, into the same expense account that is used to record the expense of the Health insurance when you prepare a check for the insurance Company.
  In the **Liability account** field, select an *account* that is *Expense* type. As stated above, select the same expense account that you use to record the health insurance Company's checks (when you pay them), which is the *Group Health expense*. That's the account that the expense will accumulate as shown in figure 4.9.
  Click the **Next** button to continue.

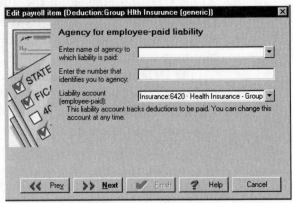

Figure 4.9 shows the expense account for the Group Health insurance program.

- The **Tax tracking type** screen is where you determine whether to track this Item for tax purposes.
Note: This is a tax issue, if your are not sure whether a certain deduction should be tracked, please consult with your accountant.

    A deduction for a group health program most likely should <u>not</u> be tracked. However, your program may qualify to be tracked for tax purposes. There are many different variations of plans. Please consult with your accountant regarding your plan.
    If your plan <u>does</u> qualify, that means employees will enjoy a reduction of their tax liability because the tax calculation will take place after the deduction. If this is the case for your plan, the proper selection is the **Fringe Benefits** option.
    Click the **Next** button to continue.
- Based on the selection of the previous step, this new screen comes with default settings. Make sure you set the Payroll Items according to government regulations.
    Note: When an Item has a checkmark (√) next to it, the deduction will be deducted <u>first</u> and then the tax will calculate. An Item without a checkmark will calculate the tax <u>first</u> and then the deduction will take place.
    Note: Please, consult with your accountant for the proper position of the Payroll Items in this screen. Also, after you set the Payroll Item position in

this screen, run a sample payroll for an employee and check the figures to insure accuracy.
Click at the **Next** button to continue.

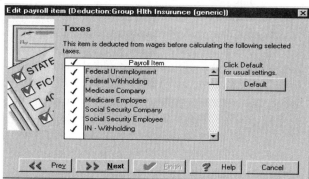

Figure 4.10 shows a possible setting for the **Group Health** deduction Payroll Item.

- At the **Calculate based on the quantity** screen, do not click at the **Based on Quantity** field.
  Click at the **Next** button to continue.
- At the **Default rate and limit** screen, type the *rate* or *amount* that you want to deduct for group health and leave the **Annual Limit** field blank.
  Click at the **Finish** button to complete this Item.
  You only need to create one Group Health Item.

e. The next Item type is the **Company Contribution**. This type allows you to setup company-paid benefits such as health or life insurance, 401(k) plans and other plans.
Company contributions are an expense to the company and have no effect on the gross pay amount but they will decrease the net pay if you set them up as taxable items in the **Edit payroll item** screen.

Gross Pay Calculation Sequence:
It is important that we mention the sequence that QuickBooks® uses to calculates the gross pay.
The gross pay calculation sequence is as follows:
Gross = salary + hourly wages + commissions + company contributions + additions

The order in which you enter payroll items in the *Employee Template* and on the *Preview Paycheck* window (as we'll see later), can affect the amount QuickBooks® calculates for each item and for taxes.

The position of an *addition* to the gross pay affects how QuickBooks® calculates *deductions* that are based on a percent of the gross.

For example, if an employee has a salary of $1000, an addition to the gross of $200 and a 5% deduction from the gross, here is how it will calculate:

1. If the <u>addition</u> is entered ahead of the deduction, it will calculate the 5% deduction on a gross of $ 1200 ($1000 + 200 = 1200). The net amount will be **$1140** (1200 – 5%).
2. If the <u>deduction</u> is entered ahead of the addition, it will calculate the 5% on a gross of $1000 and the net will be **$1150** (1000 – 5% + 200).

f. The next Item type is the **Federal Tax**. This type allows you to set the following:
   - The default positions for the type of *accounts* that you want to be used in the payroll
   - And the government agency *names* to be used on the checks to pay the taxes.

QuickBooks® creates the **Federal Tax** items when the user selects the Preferences and clicks at the **Full payroll features** option. They are in your Payroll Items list already. However, you need to **edit** each one of them in order that you set the *names* of the Federal agencies and the *accounts* from the chart-of-account as stated above.

To edit a **Federal Tax** Item, select the following:
- The **Lists** at the menu bar.
- The **Payroll Items** from the drop down menu.
- At the Payroll Items screen, scroll down to the **Federal Tax** type, as shown in figure 4.11 and select one Item at a time, highlight it, and click twice on it. You are ready to begin the <u>editing</u> of the Item.

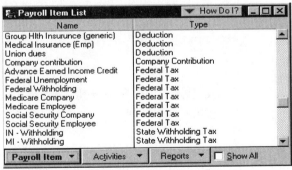

Figure 4.11 shows the section of **Federal Tax** type Items.

161

Federal Unemployment Item:
- Click at the **Next** button (because the name is just a default name).
- At the next screen, type a name at the **Enter name of agency** field. This name will print on the check at the "Pay to the order of" field.
  Note: The names for all the Federal Tax items most likely ought to be your Bank's name (that's where you are paying the company and employee withholding taxes).
- At the **Liability account** field, select an account that is *liability* type with a name like FUTA Payable.
  Note: You may want to create a group of liability accounts to use with the Federal and State and Local Payroll Items as shown in figure 4.12.

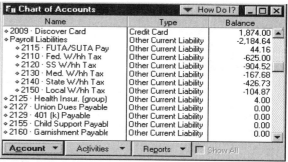

Figure 4.12 shows the **payroll liability accounts** in the chart-of-accounts.

- At the **Expense account** field select an expense type *account* such as Payroll Tax Expense. This account will receive the expense of the FUTA (the same account should also be used for the SUTA and the company portion of Social Security and Medicare). You only need one Payroll Tax Expense account. Click at the Next button to continue.
- At the next field, select the FUTA rate. The lower rate of .08% may possibly apply to your company. *Please consult with your accountant if your are not sure.*
- At the **Taxable compensation** screen, make sure all the payroll Items have a checkmark (√), click at the **Next** button to continue and at the **Finish** button to complete the Item.

For Federal, Social Security and Medicare Items, continue the same steps outlined for the Federal Unemployment Item. Remember to select the proper accounts for these Items to default into:

Item Name to choose:	Accounts you'll need to accumulate the liability:	Expense Acc:
Federal Withholding	Fed. Withholding payable	N/A
Medicare Comp.	Med. Withholding payable	Payroll Expense
Social Security Comp.	Soc. Sec. Withhdg. payable	Payroll Expense

g. The next Item type is the **State Tax**. This type allows you to create State Payroll Items for withholding taxes as well to create default positions for the type of accounts and the agency names to be used on the checks to pay the State and Local governments.

To create a **State Tax** Item, select the following:
- The **Lists** at the menu bar.
- The **Payroll Items**.
- At the Payroll Items List screen, select the **Payroll Item** button and the **New** option from the menu.
- Click at the **State Tax** Item and the **Next** button to continue.
- At the State tax screen, select the proper **State** from the list, and next click at the appropriate button that is applicable to your State such as shown in figure 4.13.
Click at the **Next** button.

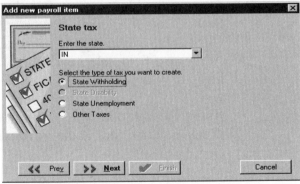

Figure 4.13 shows the **State Tax** screen.

163

- At the **Agency for employee-paid liability** screen, type the *name* of the agency where the check will be sent, the State company ID number and select an account where the State withholding tax will accumulate until you print the check for the State. The account must be a *current liability* with a name such as **State Withholding Tax Payable** as shown in figure 4.14. Click at the **Next** button to continue.

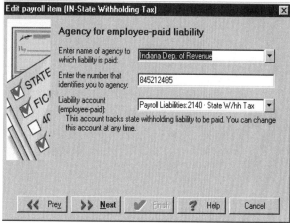

Figure 4.14 shows the **Agency for employee-paid liability** screen.

- At the next screen, all the taxable Items in the center of the screen should have a checkmark (√) next to them. Click at the **Next** and the **Finish** buttons to complete the Item.

h. The next Payroll Item type is the **Other Tax**. This type of Item, allows you to create Local type Items for withholding of local taxes and create the default positions for the following:
   - The type of *accounts* you want to be used in the payroll, and
   - The agency *names* to be used on the checks to pay the tax to Local governments.

To create an **Other Tax** Item, select the following:
- The **Lists** at the menu bar.
- The **Payroll Items** from the drop down menu.
- At the Payroll Items List screen, select the **Payroll Item** button and the **New** option from the menu.
- Click at the **Other Tax** Item and the **Next** button to continue.

164

- At the **Other Tax** screen, select the proper **Name** from the list, and click at the **Next** button to continue.
- At the **Agency for employee-paid liability** screen, type the *name* of the agency, where the check will be sent, the company ID number, and select the account for the **Other tax** to accumulate in until you print the check. The account must be a *current liability* with a name such as **Local Withholding Tax Payable** as it's shown in figure 4.14. Click at the **Next** button to continue.
- At the **Taxable compensation** screen, all the taxable Items in the center of the screen should have a checkmark (√) next to them. Click at the **Next** and the **Finish** buttons to complete the Item.

Enter Employees

*P*urpose: To create employee templates that will contain the information used for tracking time and for the payroll.

To create a new **employee** template in QuickBooks®, select the following:
1. The **Lists** at the menu bar.
2. The **Employees** option from the drop down menu.
3. At the **Employee List** screen, click at the **Employee** button once, and select the **New** option
4. At the **New Employee** screen, type the following:
   - A *salutation* such as Mr., Ms., etc.
   - *First* name
   - *Last* name
   - *Address, City, State, and Zip code.*
5. Next, click at the **Payroll Info** tab.
6. At the **Earnings** window, select the Payroll Items from the list that are *salaried* and or *hourly* type only:
   - You may select salaried or hourly or both
   - If the employee is hourly, you may select up to eight (8) hourly type Items such as Regular, Over Time1 (OT), OT2, OT3, etc.

- Under the **Hour/Annual Rate** field type either, the **annual** salary amount if employee is salaried or an **hourly** rate if hourly.
7. At the **Additions, Deductions and Company Contributions** window, select from the Payroll Item list Items that are Commission, Additions, Deductions type and etc., as it's shown on figure 4.15.

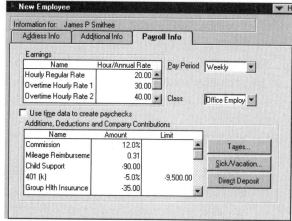

Figure 4.15 shows the completed **Payroll Info** tab for James P. Smithee.

8. At the **Pay Period** field, select the type of pay such as Weekly, Monthly, etc.
9. At the **Class** field, select a class from the list for the purpose of tracking the payroll wages into a report by Department (as we have described at Class section on Chapter 1) or into a report for the Workman's Compensation auditor. Next click at the **Taxes** button.
10. In the **Taxes** screen and:
    At the **Federal** tab, select the following:
    - The Filing status
    - Type the number of allowances
    - Type any extra withholdings. QuickBooks® can withhold extra, apart from the tax tables
    - Make sure that Social Sec., FUTA and Medicare field are all selected with a (√)

    At the **State** tab, select the following:
    - The state at the **State Worked** field
    - The state at the **State Lived** field
    - Type the number of allowances
    - Select the proper Filing Status from the list available to the State

At the **Other** tab, select the following:
- The *name* of the local taxing authority at the Name field (the name of the local tax authority is the name you create via the Other Tax selection at the Payroll Items as figure 4.1 shows).
- Type the *name* of the locality at the **Print on W-2 as** field.
- Type the *percent* of the tax your locality requires.
- Click at the **OK** button to save your selections.

11. Next click at the **Sick/Vacation** button to set the sick and vacation requirements so QuickBooks® can track sick and vacation time that accrues for this employee.
    To set the Sick and Vacation requirements, type and or select the following information:
    - Select the **Every pay period** field if you want QuickBooks® to track time each time you do payroll
    - Type the **rate** that you want QuickBooks® to accrue at the **Hours accrue per accrual period** field is **.77** hours per week if the maximum of sick pay was 40 hrs. and assuming you do payroll every week (based on 52 weeks, 40 / 52 = .77). Likewise the rate is **1.54** hours if the maximum for vacation pay was 80 hrs. and the payroll is done weekly. Then, 80 / 52 = 1.54, as shown in figure 4.16.
    - Next, enter the **maximum** hours of Sick and Vacation pay as figure 4.16 shows.
    - Click at the **OK** button. At the next screen, click at the **OK** button again to save the employee information.

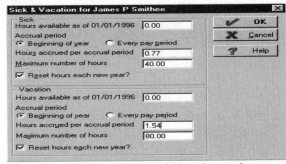

Figure 4.16 shows the sick and Vacation settings.

Tracking Employee Time

𝒫urpose: To keep track of employee time that can be used in time reports, to bill customers with billable time and to do the payroll. The time tracking ability is found only with the QuickBooks Pro® version.

Time can be tracked through the **Single Activity** function and the **Weekly Timesheet**.
- A. The **Single Activity** allows you to track time one employee, one day at a time
- B. The **Weekly Timesheet** allows you to keep track of time for the entire week, one employee at a time.

Time through the **Single Activity,** can be tracked either via the *automatic* mode by using the stopwatch or *manually* by typing the time. The stopwatch works only for the current date and not for a past or forward date.

To keep track of employee time using the **Single Activity**, select the following:
1. The **Activities** at the menu bar (in QuickBooks® 2000, select **Employees**).
2. Next, select the **Time Tracking** function from the drop down menu.
3. The **Time/Enter Single Activity** option.
4. At the **Time/Enter Single Activity** screen, select the following:
    a. An employee *name* from the list
    b. If you want to keep track of time by Job, you can select the *name* of a Customer:Job at the **Customer:Job** field. If you do not select a name at this field, there would be no reporting of employee time available by Job. Once you select a customer or job name you <u>must</u> select also a Service type Item in the next step.
    c. At the **Service Item** field, select an Item from the list that is *Service* type
    d. Next, highlight the field under **Duration**, on the stopwatch, and type manually the amount of hours the employee has worked
    e. Next, select the **Billable** field on the upper right and place a checkmark (√) to indicate if the time is billable. Without a checkmark (√) you cannot bill for employee time.
    Note: The **Billable** field, allows you to bill time to a customer with whom you have an agreement that you will bill for time spent working at the customer's project.
    f. Select the type of *pay* you will be paying the employee at the **Payroll Item** field.
    g. At the **Class** field select a Class from the list in order to track the time into a report by Department or for the Workman's Compensation auditor.
    h. Next, click at the **OK** or the **Next** button.

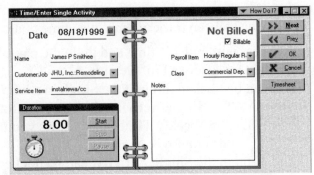

Figure 4.17 shows the completed **Single Activity** screen and the time tracked for James Smithee.

To keep track of time using the Stopwatch, you can follow all the steps outlined in the previous example except, in step 4d, click at the **Start** button on the Stopwatch and it will begin counting of the time as its shown of figure 4.18. At the end of the day, or at another time, the employee can stop the watch by clicking at the **Stop** button on the Stopwatch. Next, click at the **OK** button to complete recording time or you may click at the **Next** button to record time for another employee or the same employee.

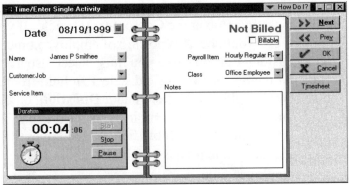

Figure 4.18 shows the **Stopwatch** active.

To use the stopwatch, an employee has to work in the Office and be able to start and stop the Stopwatch, as it is needed.

To keep track of employee time using the **Weekly Timesheet**, select the following:

1. The **Activities** at the menu bar.
2. The **Time Tracking** function, from the drop down menu.
3. The **Use Weekly Timesheet** option.
4. At the **Weekly Timesheet** screen, select the following:
   i. An employee *name* from the list at the **Name** field, at the top of the screen
   j. If you want to keep track of time by Job, you need to select the *name* of a customer or job at the **Customer:Job** field. If you do not select a name at this field, there would be no reporting available of employee time by Job. Once you select a customer or job name you <u>must</u> select also a Service type Item.
   k. At the **Service Item** field, select an Item from the list that is *Service* type
   l. Select the type of *pay* you will be paying the employee at the **Payroll Item** field.
   m. At the **Class** field select a Class from the list in order to track the time into a report by Department or for the Workman's Compensation auditor.
   n. Next, under each *day*, type the *time* the employee has spent working each day of the week.
   o. At the far right, there is the **Icon** that provides the ability to make time **billable**. To make time billable, simply leave the Icon intact. If you click once on it, it places an X and thus removes the billable option.
   p. Next, click at the **OK** button at the lower section of the screen to complete the recording or time.

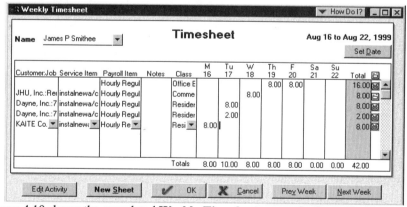

Figure 4.19 shows the completed **Weekly Timesheet** screen for James Smithee.

Notice on figure 4.19 that all the icons (to the right) have an X, except one. The one without the X, is the one we designated as billable when we tracked the employee's time.

170

Time Reports

Employee time that is recorded through the Time Tracking function can be reported by Name, Job or by Item.

To print the **time** reports, select the following:
1. The **Reports** at the menu bar (in QuickBooks® 2000, select **Reports**).
2. The **Project Reports** from the drop down menu (in QuickBooks® 2000, select **Jobs & Time**).
3. Next, select the **Time** report you want to see:
   a. **By Name**, creates a report with emphasis on the employee name
   b. **By Job**, creates a report with emphasis on the job name
   c. **By Item**, creates a report that emphasizes the service item the employee has provided.
4. After you select an option, i.e. By Name, you may add important other information to the report by clicking at the **Customize** button (on the upper left corner of the report screen). By clicking at **Billed**, **Unbilled** and **Not Billable** buttons, you will add columns that will inform you if time is billable and whether you have billed the customer for it, as shown on figure 4.20.

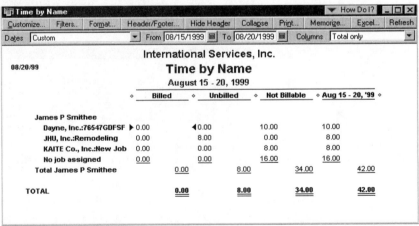

Figure 4.20 shows the time by *Name* report with additional information.

171

Invoicing Billable Time

Any employee time that you designate as billable through time tracking, can be invoiced at anytime.

To invoice for **billable** time, select the following:
1. The **Activities** at the menu bar (in QuickBooks® 2000, select **Customers**).
2. The **Create Invoice** function.
3. Select the *customer* that you wish to bill at the **Customer:Job** field
4. Next, select the **Time/Costs** button
5. At the **Choose Billable Time and Costs** screen, select the **Time** tab and place a checkmark (√) by clicking. Notice that the *Service* item appears with its default *rate*, and the *hours* come from the time tracking.

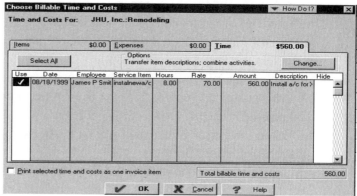

Figure 4.21 shows the billable time.

6. Next, select the **OK** button and instantly QuickBooks® creates an Invoice.

The Time Tracking is a useful function because it allows you to know where the employees have spent their time. It is extremely useful for tracking time by job.

Pay Employees

*P*urpose: To create paychecks for the company employees.

In order to do payroll you do not have to use the Time Tracking function. The time that an employee has worked for a given period can be typed manually. QuickBooks® multiplies the time with the employee's rate.

If you do use the *Time Tracking* function, there is no need to type the time manually because the employee time automatically goes into the payroll.

Employee time in the payroll from either the *Time Tracking* or from *manual* input can be edited.

To prepare **paychecks** for the company employees, select the following:
1. The **Activities** at the menu bar (in QuickBooks® 2000, select **Employees**).
2. The **Payroll** function from the drop down menu
3. Next, select the **Pay Employees** option.
4. At the **Select Employees to Pay** screen, proceed with the following:
   a. Click to place a checkmark (√) at the **To be printed** field, so that paychecks will be printed.
   b. Click at the **Enter hours and preview checks** field so that you can preview the paychecks before printing.
   c. Make certain that the date at the **Pay Period Ends** field is correct. That's the date of the end of the pay period.
   Note: If this field keeps coming up with an old date, that's an indication that you did not pay an employee as of that date (the defaulting data). To correct this situation, make the employee Inactive. Inactive employees, are removed from the Employee list and you can make them active by removing the checkmark from the **Employee is inactive** field at the employee Template.
   d. Click under the √ column to create a paycheck for an employee. Click next to only one or multiple employees.
   e. Next, click at the **Create** button to create paychecks as shown on figure 4.22.

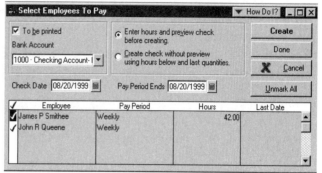

Figure 4.22 shows the **Select Employee To Pay** screen.

5. Next, in the **Preview Paycheck** screen, proceed with previewing the following:
   a. The hours (that came from Time tracking)
   b. The **Customer:Job** column and ensure that you have the proper Job names. <u>This is the field that reports the payroll $ costs into the job reports</u>. Through the **Customer:Job** field, you can change into another name or you can add a name in this column to track the cost into a Job report if an entry came without a name.
   c. Type any *amounts* in the **Quantity** column for such thing as commission, miles driven, etc. as shown on figure 4.23.

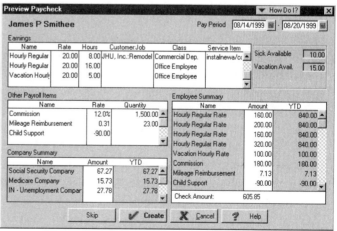

Figure 4.23 shows the **Preview Paycheck** screen.

174

d. If you need to pay vacation, place your cursor at the very last spot under the **Earnings** column (upper left), and select the **Vacation** payroll item by clicking. Under **Hours** column, type the vacation hours you want to pay. After you type the vacation hours, QuickBooks® automatically deducts the hours from the **Vacation Available** field.
6. When you are ready to create the paycheck, select the **Create** button and QuickBooks® provides the next employee. If there is no other employee, it creates the paycheck and stops.
7. At the next screen, click the **Done** button to close the function.

Print Paychecks

To **print** paychecks, select the following:
1. The **File** at the menu bar.
2. The **Print Forms** option from the drop down menu.
3. And select the **Print Paychecks** option.
4. At the **Select Paychecks** screen, place a checkmark (√) next to each paycheck you want to print.
5. Click the **OK** button to begin the printing.

Edit Paychecks

In the event that you make a mistake while doing the payroll, QuickBooks® allows you to edit paychecks.

To **edit** a paycheck, select the following:
1. The **Lists** at the menu bar.
2. Next, select the **Chart-of-Accounts** option from the drop down menu.
3. Select the **Checking** account that you have used to for the paycheck (s), highlight the account and click twice. You are now in the account's **Register**. Next, find the paycheck you want to edit, highlight it and click at the **Edit** button once.
4. In the paycheck, click at the **Paycheck Detail** button and you are now in **the Preview Paycheck** screen (paycheck detail).
   You may *edit* the Rate, the Hours, **Customer:Job** and Class columns. When you're done, click at the **OK** button. And click again at the next **OK** button, at the paycheck screen. You have finished editing a paycheck.

Payroll Reports

After you do payroll, it is a good business practice that you print a payroll report such as the **Summary** report. Payroll reports must be kept through the calendar year for emergency purposes.

To print the **Summary** payroll report, select the following:
1. The **Reports** at the menu bar
2. The **Payroll Reports** option (in QuickBooks® 2000, select **Employees & Payroll**).
3. Next, select **Payroll Summary** from the side menu.

Pay the Liabilities and Taxes

After you do the payroll you must pay the payroll taxes. Depending on the amount of your payroll taxes, the *Internal Revenue Service* will designate the time intervals that you need to follow for depositing the payroll taxes with the designated tax depository in your area.
To pay the Federal, State and Local payroll taxes on time, make sure that you understand the time intervals that you are supposed to follow.

To pay the Federal and State payroll taxes, you must have the required federal and State tax coupons. If you do not have the coupons, please call your accountant or the appropriate Federal and State agencies.

To **pay** the **payroll taxes**, select the following:
1. The **Activities** at the menu bar (in QuickBooks® 2000, select **Employees**).
2. Next, select the **Payroll** function (in QuickBooks® 2000, select **Pay Payroll Liabilities**).
3. Next, select the **Pay Liabilities/Taxes** option.
   (For QuickBooks® 2000, you need to select the date at the **Show liabilities from** field).
4. At the **Pay Liabilities** screen, place a checkmark (√) next to the payroll item you want to pay such as the Federal, Social Security, Medicare, State and Local and etc.
5. When you have finished selecting, click at the **Create** button and QuickBooks® will create the checks to pay the Federal, State and Local governments and any other items you may have selected such as Child support, Garnishments, etc. as shown in figure 4.24.

The checks you create through the **Pay Liabilities** screen go directly into the batch print and wait for your next step. To print these checks, follow the steps that are outlined below.

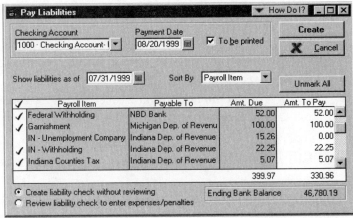

Figure 4.24 shows the Pay Liabilities screen.

To **print** the **payroll liability** checks, select the following:
1. The **File** at the menu bar
2. The **Print Forms** option from the drop down menu.
3. Next, select the **Print Checks** option.
4. Click next to each check that you want to print under the √ column.
5. Click at the **OK** button to begin the printing process.

Payroll Forms

941 Form

This form allows you to report the payroll withholdings of the Federal, Social Security and Medicare taxes that you have withheld from the employee paychecks.
Through this form, you also report the company's *share* of the Social Security and Medicare tax. It's the form you need to file with Federal government every calendar quarter, before the end of the following month.

QuickBooks® is capable of accumulating the information and printing the 941 form on plain white paper. The government accepts this form as it's printed from QuickBooks®.

To print **941 Form**, select the following:
1. The **Activities** at the menu bar (in QuickBooks® 2000, select **Employees**).
2. Next, select the **Payroll** function (in QuickBooks® 2000, select **Process Payroll Forms**). Notice: In QuickBooks® (and QuickBooks Pro®) 2000 if you did not select to subscribe to the Intuits payroll Tax service, you will not have available the ability to print the payroll 941 and 940 forms.
3. The **Process Form 941** option (in QuickBooks® 2000, select the **941** form).
4. At the **Form 941** screen as shown in figure 4.25, you have four options. Select one and click at the **OK** button to proceed. The four options are the following:
   a. The **Create Form 941** option allows you to create a **new** 941 form. If you select this option, you will delete the previous quarter 941 form and QuickBooks® will create a new 941, for the current quarter.
   Next select a quarter-end *date* (the date of the quarter you want to create, this starts the process of completing the form) and click at the **OK** button to proceed.
   b. The **Edit form 941** option, allows you to edit a previous form
   c. The **Print form 941** option, allows you to print the form that is currently available
   d. The **Preview form 941** option, allows you to preview last quarter's 941 form
5. At the Form 941 screen select the State code, and select the **Next** button
6. At the **Line 1** field, enter the *number* of employees that have worked for your company during the quarter
7. At the next three screens you have the opportunity to adjust wages, the Social Security and Medicare taxes. Under normal circumstances, you should not have a need to adjust any of the above. However, if you do, click at the Yes button and

proceed accordingly in each screen. The information you adjust on this form, does not adjust the payroll.
8. At the last screen, you have the option of printing the form or previewing it now and printing it later.

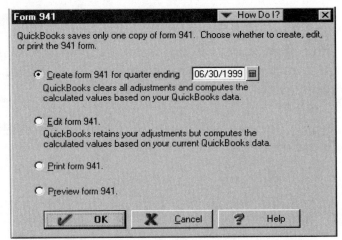

Figure 4.25 shows the **Form 941** options screen.

You can print the 941 form through a Dot matrix or a laser printer. The Dot matrix usually takes much longer then the laser. You can print on plain white paper because QuickBooks® is able to print the form alphanumeric characters in the same way as the original Internal Revenue Service form.

940 Form

This form allows you to report the FUTA (Federal Unemployment Tax Act) tax which is a company expense. It's the form you need to file with Federal government at the end of the calendar year, but before the end of January 31st.

QuickBooks® is capable of accumulating the information and printing this form on plain white paper. The government accepts the form as it is printed from QuickBooks®.

To print the **940 Form**, select the following:
1. The **Activities** at the menu bar (in QuickBooks® 2000, select **Employees**).

2. Next, select the **Payroll** function (in QuickBooks® 2000, select **Process Payroll Forms**). Notice: In QuickBooks® (and QuickBooks Pro®) 2000 if you did not select to subscribe to the Intuits payroll Tax service, you will not have available the ability to print the payroll 941 and 940 forms.
3. The **Process Form 940** option
4. At the **Form 940** screen, shown in figure 4.26, you have four options, select one and click at the **OK** button to proceed to the next step. The four options are the following:
   e. The **Create Form 940** option allows you to create a new 940 form.
   f. The **Edit form 940** option, allows you to edit the current year's form
   g. The **Print form 940** option, allows you to print the form that is currently available
   h. The **Preview form 940** option, allows you to preview the current year's form 940
5. At the **Form 940** screen make sure the current year appears at the **Edit/Verify** field, and click at the **Next** button.
6. At the next screen, answer the 940 filing information that is required and click at the **Next** button.
7. At the next few screens you have the opportunity to adjust Part I, II and III of the form 940. As in the 941 form, you should not have a need to adjust any of the payroll information. However, if you do, click at the Yes button and proceed accordingly in each screen. The information you adjust on this form, does not affect the payroll.
8. At the last screen, you have the option of printing the form or previewing it and printing it at a later time.

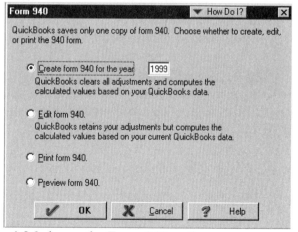

Figure 4.26 shows the Form 940 options screen.

You can print the 940 form through a Dot matrix or a laser printer just as form 941.

180

W-2 Form

It's the form you need to provide to every one of your employees at the end of each calendar year. The W-2 forms must be printed and delivered to employees before the end of January 31st.

QuickBooks® has the ability to accumulate the payroll information and print the W-2s. The W-2 forms that can be purchased from Intuit, Inc.
To **print W-2 forms**, select the following:
1. The **Activities** at the menu bar (in QuickBooks® 2000, select **Employees**).
2. Next, select the **Payroll** function (in QuickBooks® 2000, select **Process W-2s**).
3. The **Process W-2s** option.
4. At the **Process W-2s** screen as shown in figure 4.27,
    a. Make sure that you have the proper *year* at the **Year** field.
    b. Click once next to each employee to place a checkmark (√).
    c. Next, click at the **Review W-2** button to review each W-2 form. When you have finished all the forms, click at the **OK** button.
    d. Next, *load* your printer with the actual **W-2** forms.
    e. Next, if you are printing through a Dot matrix printer, try printing one employee's data so that you may make any necessary adjustments to the form. After you adjust the forms on the printer, you can begin the printing.
    f. Click at the **Print W-2s** button.
    g. Next, click the **Print W-3** button to print the W-3 form. This is the form that accompanies the W-2s. The W-3 can be printed in plain white paper because you will need to either type or write it by hand on the final form that you send to the Federal government. All you need is the information from the system.
    h. Next, click the **Done** button to end the process.

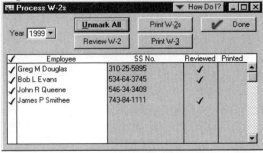

Figure 4.27 shows the **Process W-2s** screen.

Chapter 5

Other Functions & Misc. Information

*T*he purpose of this chapter is to help you learn how to make journal entries and use the bank reconciliation. It'll also help you understand how to use other important functions that are available in QuickBooks®.

In this chapter we'll examine the following QuickBooks® functions and misc. topics:

- **Journal Entries**
- **Bank Account Reconciliation**
- **Write Letters**
- **Accountant's Review**
- **Timer Activities**
- **Export Addresses & Lists**
  (Mail Merge)
- **Export Files**
- **Back Up & Restore**
- **Customizing Forms**
- **Budgeting**
- **Year-end Activity**
- **Keyboard Shortcuts**
- **Glossary**

Make Journal Entries

*P*urpose: To record special transactions such as year-end (and other times) *adjustments* to account balances, the *depreciation* transaction, to adjust for *prepayments*, to record *Bank charges*, *interest* income earned from interest bearing accounts and etc. Also, through the Make Journal Entries function, you may record Bank loans and telephone transfer of funds.

Via the **Journal Entries** function, *beginning account balances can be entered* for a new company either at the beginning or at a later time (if you have been working in QuickBooks® for a while without the beginning balances, you may enter them as of the end of month).

Using the **Make Journal Entry** function, we'll provide a number of example transactions that you may adapt for your own company's needs.

To record a **journal entry**, select the following:
1. The **Activities** at the menu bar (in QuickBooks® 2000, select **Company** or **Banking**).
2. The **Make Journal Entry** option at the drop down menu
3. Select or type the *date* of the transaction at the **Date** field.
4. Type in an ID or number for each transaction at the **Entry No**. This is an optional entry. You may type alpha and or numeric characters. This field, appear on reports.
5. Next, proceed by selecting an *account* under the **Account** field
6. Next, type an *amount* either under the **Debit** or the **Credit** fields.
   <u>Below, we provide a few useful examples that would provide direction as to where the amounts ought to be entered.</u>

Example #1: Recording the **Depreciation** transaction.
To obtain the figures you'll need to record depreciation, you must work with your accountant. Your accountant will figure out a depreciation schedule for each fixed asset. You must divide each annual amount on the depreciation schedule by twelve (12), the months in the year, to arrive at monthly figures.
Business tip: This transaction must be recorded at the end of each month, every month, so that your financial statements will be complete and therefore, accurate, throughout the year.

To record depreciation, you must create the following accounts:
1. An *expense* type account that you must name **Depreciation Expense**
2. A *fixed asset* type account(s) you must name **Accumulated Depreciation** (consider abbreviating this account(s) as **AD**). You must create one AD account for each fixed asset that exists in your books.

Note: For a thorough understanding of the financial **categories**, the **accounts** in the chart-of-accounts, and how the process of **accounting** works, we highly recommend the "Understanding Accounting" Videotape developed by the author of this book. In this tape, the author uses practical examples he presents in a simple, layman's language. To order, call 219-482-3399 or Fax 219-471-5302.

Let's say your company has a two year-old (2) building it uses as office & storage combination, and a ten (10) month-old piece of machinery it uses in the shop.

	Cost	Depr./Year	Depr./Month
The cost of building in the books	$75000	2nd yr., $16073	$1339
The cost of equipment in the books	$36000	1st yr., $9000	$750

To record the Depreciation transaction properly, follow the example below:

	DR	CR
Depreciation expense account	$2089	
AD-Building account		1339
AD-Equipment account		750

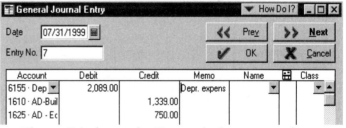

Figure 5.1 shows the **Depreciation** transaction.

Example #2: Recording an adjusting entry for the **prepayment** of an expense.

If you decide to prepay an *expense* such as the rent, insurance, interest, etc. for the entire quarter or year, the entire amount of the prepayment, at the time you print or hand-write the check, must be recorded into an *asset* type account. You must create this account in the chart-of-accounts and name it **Prepaid Expenses**.

After you record a *prepayment*, you need to follow through every month, with recording an adjusting entry via the *Make Journal Entries* function. Through this entry, you will be adjusting the *asset* account that was used to record the prepayment (when the check was printed) and an *expense* account. Actually, you'll be decreasing the asset account's balance and increasing the expense account.

For example, let's say you have issued a check sometime at the begin of the year, through the *Write Checks* function in order to prepay the insurance for this entire year. The amount of the check was for $1800.
For every prepayment, you need to record an **adjusting** entry. These adjusting entries are accomplished through the **Make Journal Entry** function.
To record the transaction of the adjusting entry at the end of a month, follow the example below:

	DR	**CR**
Insurance expense	$150	
Prepaid expenses		150

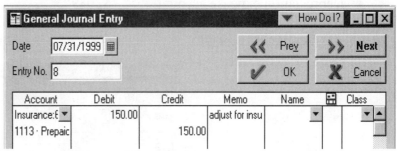

Figure 5.2 shows the adjusting entry for the **prepayment** of the insurance.

Example #3: Recording a long-term Bank **loan**.
To record a loan, you first have to create a *Long-Term* type account in the chart-of-accounts with a name such as **LT Loan – Bank X**. After you create the account, you may record this transaction via the *Make Journal Entries* function.

For example, let's say that your company has borrowed from the Bank $30000 to purchase a truck. The loan is to be paid back in three (3) years, at the rate of 9.5%.

To record the transaction of **borrowing** from the bank, follow the example below:

	DR	CR
Checking account	$30000	
LT loan – Truck account		30000

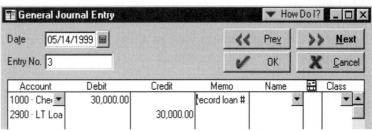

Figure 5.3 shows the recording of the **Bank loan**.

Example #4: If you are in the construction business and you build spec homes, the amounts you may need to borrow from the Bank to finish each specs house can be recorded in the same way as example # 3 above.

For example, let's say that you are approved by the Bank to borrow $110,000 to build a spec house that you hope you'll sell soon. And at this point, you've received from the Bank the first advance of $15,000 that you can use to pay your vendors and suppliers. To record the advancing of the funds (the loan), you must increase a liability account that should be a **Long-Term Liability** type account (if the loan has a maturity date of more then twelve months) and a **Current Liability** type (if the loan is due in twelve or less months).

To record the **loan** from the bank, use the same accounting as in the example below:

	DR	CR
Checking account	$15000	
LT Loan – Spec House X		15000

Business tip: Create a Long-Term Liability account in the chart-of-accounts for each spec house.

Example #5: Recording **interest** earned from an interest bearing account at the Bank such as Savings or other type of account.

After you create an account in the chart-of-accounts with the name **Interest Income** and type *Other Income*, you may record this transaction via the *Make Journal Entries* function.

Let's say that the statement you've received from the Bank says that your checking account (or a savings account) has earned $25 in interest during the past month.

To record the interest earned, use the same accounting as in the example below:

	DR	**CR**
Checking account	$25	
Interest Income		25

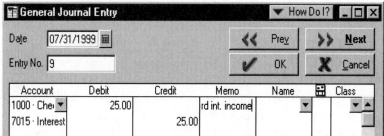

Figure 5.4 shows the transaction of recording **interest income**.

Bank Account Reconciliation

*P*urpose: To ensure that there are no errors or fraudulent transactions in the Bank (Checking) account's activity and to verify the accuracy of the account balance.

At the end of each month, you receive the statement from the Bank that shows the monthly activity of your account. This monthly activity must be compared against the information you have in the system.

The process of reconciling (or comparing) the bank statement against your system's information is accomplished through the **Reconcile** function. This same function is also used to reconcile *Credit Card* statements.

The desired objective of the reconciliation process is that you end-up with a zero (0) difference at the **Difference** field. That will indicate that the *ending statement balance* agrees with the system's activity.

To **Reconcile** the bank statement, select the following:
1. The **Activities** from the menu (in QuickBooks® 2000, select **Banking**).
2. The **Reconcile** function from the drop down menu.
3. At the **Account to Reconcile** field, in the **Reconcile** window, select the bank account **name** from the list.
4. Next, enter the *ending balance* amount that appears on the statement by typing it in the **Ending Balance** field.
5. Enter by typing the following:
   a. Type the *charges* amount in the **Service Charge** field
   b. Select the *date* that the charge occurred such as a month end, in the **Date** field.
   c. Next, select an *account* in the **Account** field from the chart-of-accounts list to record the expense of the charge. The appropriate account would be the *Bank Charges,* which is an expense type account. The account you select, becomes the default account in the future.
   Let's say there is a $10 service charge in the past month's statement, as soon as the amount is typed; the following accounting is completed in the background:

	DR	CR
Bank Charges account	$10	
Checking account		10

d. Next, enter by typing any *interest* the account may have earned in the **Interest Earned** field.
e. Type or select the date interest occurred in the **Date** field. Next, interest earned at the bank, ought to be recorded as income in your books. Therefore, the account to use should be the *Interest Income*. This account must be type *Other Income*.
For example, let's say your account at the bank is an interest bearing account and it has earned $25 during the past month. When you type the amount of the interest in the *Interest Earned* field, the following accounting is completed in the background:

	DR	CR
Checking account	$25	
Interest income account		25

6. Next, begin the process of reconciling one *transaction* at-a-time and one *section* at-a-time:
Deposits and Other Credits (the upper section) and the **Checks and Payments** section (lower section):
   a. View the following fields and make sure that what's in the system agrees with what's on the statement: **Date, Chk No., Payee, Amount.**
   b. If the information in these fields *does* agree on both, the statement and the system, then place a *checkmark* (√) next to the transaction by clicking once. Placing a *checkmark* next to a transaction, means the transaction is OK (no problem). It also tells the system that the transaction has **cleared**.
   c. If the transaction appears on both, the statement and in the system, but the amount may be different on the statement. Then you need to find the actual check and verify the amount. It could be that the Bank has paid out the wrong amount. Then you need to contact the Bank and report the error.
   d. If a transaction appears in your system, but it doesn't appear on the statement, that indicates that the transaction did not clear the bank's process. It is uncleared. <u>Do not place a check mark next to it</u>. The absence of a check mark, tells QuickBooks® that the transaction has not been cleared yet.
   *Note: Transactions that haven't cleared (without a check mark) will appear again during the next month's reconciliation. On the other hand, transactions that have been cleared (with a check mark) will be removed from the **Reconcile** function once you finish the process and click the DONE button.*
   e. If you end-up with a difference that is other then zero, it could mean one of three

possible scenarios:

Type	Detail	Action
Error	Bank paid the wrong amount[1]	Notify the bank
Fraud	Unauthorized check[2]	Notify management
No record	Manual check not recorded	Record via **Write Checks**

[1]. Note: An error could be a check that the Bank has paid the wrong amount simply by mistake. It could also be that you have issued as amount $X, but it was intercepted before it reach the intended payee, and the amount was changed to $Y. Furthermore, the Bank has cashed (based on actual experience) the check with the new amount. Now, you are in the process of reconciling and you notice that a particular check has two amounts: Amount $X in your system, and $Y on the statement. So, an **error** type could be either: An honest mistake or fraud.

[2]. Note: Usually an unauthorized check will clear the Bank but there would be no record of it in the system just like a manual check.

Please remember: If you end up with a difference that is other then zero, do not click the DONE button. You must click the LEAVE button (Cancel in QuickBooks v4 and v5). The Leave button, tells QuickBooks® to remember all the work you have done so you do not have to repeat your effort. Later, when you find the reason of the difference and return to the Reconcile function, you can continue where you left off.

f. In the event of a *Bank error* or *fraud*, you need to communicate with the Bank as soon as possible and especially with fraud. After you resolve your difference with the bank, you may select the **Reconcile** function again and continue the reconciliation process as described in step **b** above. Now, you should have a difference of zero (0). Please see figure 5.5.

Business tip: In the event of fraud, Banks usually do not assume any liability past thirty (30) days from the date of the statement.

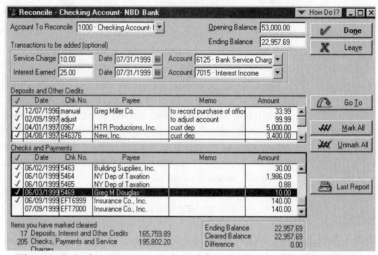

Figure 5.5 shows a completed bank reconciliation process.

g. When you finish the reconciliation process with a zero (0) difference, and then click on the **DONE** button. QuickBooks® clears all marked items in the account register and also, removes them form the Reconcile function.

h. Next, appears the **Reconciliation Complete** screen. In this step, you have three options that pertain to the type of report you wish to print: None, Summary and Full report. The Summary lists only total. The Full report shows all cleared and uncleared transactions and the totals. Select one and click the **OK** button to print the report.

Note: The amount in the **Opening Balance** field can only be changed via the account's **register**. If you change any previously recorded and cleared (reconciled) transaction in the register, the change will affect the *Opening Balance* field in the *Reconcile* function.

Write Letters

*P*urpose: To create letters that can be used to communicate with Customers, Vendors and Employees.

The **Write Letters** function in QuickBooks®, interfaces with the Microsoft® Word automatically and allows you to either use a collection of prepared letters or to customize a new letter to fit a particular purpose. When you use the prepared letters function, you are still allowed to edit the letters before you complete them.

To use the **Write Letters** function, select:
1. The **Activities** at the menu bar (in QuickBooks® 2000, select **Company**).
2. Next, select the **Write Letter** option
3. At the Write Letters screen, you have three options. Select the one you want to use by clicking:
    a. The **Prepare a Collection Letter** option is for amounts owed to your company that are over-due. It allows you to set the criteria that will identify by:
        - The *Customer* or the *Job*
        - Customers or Jobs that are *Active*, *Inactive* or Both
        - The *overdue* category by: 1 day or more, 31 days or more, etc.

    b. The **Prepare Another Type Letter** option, is designed for letters that address other needs such as: Birthdays, credit apps, bounced checks, thank you, etc. and general business correspondence. It allows you to identify by:
        - Customers
        - Vendors
        - Employees
        - Active, Inactive or Both status
        - An entire name list or selectively

    c. The **Design QuickBooks® Letters** option is to allow you to create a letter from scratch, to convert an existing Word document into a QuickBooks® letter, to edit an existing document and to perform misc. functions such as deleting, renaming, duplicating, etc.

It allows you to set criteria for:
- Vendors, addressing disputes, credit requests, payments, etc.
- Employees, addressing payroll and general business issues
- Customers, addressing collection or other general business issues

4. After you make a selection, at the next screen select the name (s), and click the **Next** button.
5. Next, select the type of letter you want to send.
6. Next, Type the name and title of the person that will be signing these letters. The name and title will appear at the bottom of each letter.
7. To create the letters, click at the **Create Letters** button.

Accountant's Review

*P*urpose: To create a special copy of your data in a disk that you can exchange with your accountant.

At the end of a quarter or year, you may need to send a copy of your data in a disk, to your accountant so that they can make adjustments to your data or for year-end tax work. In the past, this type of transferring of data between the user and the accountant was accomplished through the *Back up* and the *Restore* functions. But the restore function when it's used writes over the previous data found in the system and that results in the loss of data. In order to avoid the loss of data, the user had to *wait* without performing any work, until the accountant had returned the disk back with the adjustments. This process had resulted in lost time.

The **Accountant's Review** function has changed all that. Now, you do not have to wait for the accountant's copy to arrive back in your office, you can keep on working.

To use the **Accountant's Review** function, follow the steps outlined below:

A. User Tasks:
1. Launch QuickBooks®
2. Select the **File** at the menu bar
3. Select the **Accountant's Review** function.
4. Insert a *formatted* disk in your floppy drive **a**: or **b**: (depending on hardware configuration).
5. Select the **Create Accountant's Copy** option.
6. Accept the defaulted values by clicking the **Save** button. QuickBooks® will make a copy of your data into the floppy disk with the extension of **.qbx**.
7. After you make the copy as described above, your company's screen now will display-next to your company's file name, on the top bar, *Accountant's Copy Exists* (in parenthesis).

B. Accountant Tasks:
At the accountant's, they should perform the following steps:
1. Select the following: **File/Accountant's Review/Start Using Accountant's Copy**

2. Next, select the file from drive **A:** or **B:** with the extension **.qbx** and click the **Open** button.
3. At the next step, make sure that the file name has the extension **.qba** in the **File name** screen and at the **Save in** screen (at the top), there should appear the QuickBooks® folder as the destination of the data. Next, click the **Save** button.

   Note: At the Accountant's system, your company's data will now display in parenthesis the following message: (*Accountant's Copy*).

4. After Accountant makes the required adjustments to your data, they must prepare the data for *exporting*. To prepare the data for exporting, follow these steps:
   a. Select **File/ Accountant's Review/Export Changes for Client.**
   b. Enter a file *name* with the extension **.aif** (i.e. name.aif) at the **Save Accountant's Export** screen. Also, select the drive such **a:** or **b:** where you want the data to be saved.
      Note: The same disk received from the client <u>must</u> be used to prepare the export file.

C. User Tasks:

The purpose of this step is to *incorporate* into your company's data the changes your accountant has made in his/her copy without deleting (or writing over) the work you might have done during the same time your accountant was working on your data.

To incorporate the accountant's data, follow these steps:
1. Launch QuickBooks®
2. Print the **Trial Balance** report by selecting **Reports/Other Reports/Trial Balance** (in QuickBooks 2000, select **Reports/Accountant & Taxes/Trial Balance**). The purpose of this is to have a hard copy of your account balances before you incorporate the accountant's copy into your data.
3. Next, select the following: **File/Accountant's Review/Import Accountant's Changes** and click the **OK** button.
4. Next, QuickBooks® will remind you to Back up your data. Click the **OK** button in the next screen and insert a new disk in drive **a:** or **b:** and click on the **Save** button to proceed with the Back up.
5. Click the **OK** button at the next screen.

6. At the **Import Accountant's Changes** window, select drive **a:** or **b:** and select the file name that appears in the white screen. Click at the **Open** button.
7. Next, click the **OK** button.
8. Review the validity of your data by comparing the new Trial Balance on the screen against the one you have printed (in step # 2 above) to make sure all your work is intact and that the accountant's work is there too. Proceed with the rest of your work.

It is **very** important that you remember to take every precaution possible in order that you will not experience any loss of data. To avoid unpleasant surprises, you must backup your company data every day according to the suggested method described in the *Backup and Restore* section.

Timer Activities

𝒫urpose: To keep track of employee time that is spent working on customer projects.

Time kept in the Timer can be used in the *payroll* and to invoice customers with *billable time*. Employee time from the QuickBooks Pro Timer® can be reported through the QuickBooks Pro® reporting capability.

The QuickBooks Pro Timer® program can be very valuable to service oriented companies such as Law Firms, Accountants and any other company that needs to keep track of time by employee.

The QuickBooks Pro Timer© can be used on single workstation-where input can be done through the same computer, or through multiple workstations. In the case of multiple workstations, each employee must have QuickBooks Pro® installed on his/her computer and use the Timer to keep track of time. At the end of the week (or month), each employee will have to export his/her time into a disk, which then will be taken to the workstation where the *billing* and *payroll* functions will be taking place.

The QuickBooks Pro Timer© is a separate program that comes with QuickBooks Pro®. To begin working with it first, you need to *export* the **Lists** such as the Chart-of-accounts, Employee, Vendor, Customer, Classes, and Payroll Classes from QuickBooks Pro® and *import* them into the QuickBooks Pro Timer©. This process is accomplished via the **Export** and the **Import** functions.

Step A:
Launch QuickBooks Pro®:
The purpose of this step is to export the lists. To accomplish the exporting, select the following:
1. The **File** from the menu bar.
2. The **Timer Activities** option (in QuickBooks® 2000, select **Timer**).
3. Next, select the **Export Lists for Timer** and in the next screen click the **OK** button.
4. In the next screen, enter a file *name* and the *extension* .iif (i.e. name.iif). This list file will be exported into the QuickBooks® folder (or Directory). Next, click OK.

Step B:
- **A.** In this step, you'll start the QuickBooks Pro Timer© to import the Lists and begin keeping track of time. To accomplish this task, select the following:
  - a. At the Windows® desktop, select: **Start/Programs** click on the **QuickBooks** icon, select the **QuickBooks Pro Timer®**.
    When you are in the Timer for the first time, select the **Create New Timer File** option in the next screen
    Click the **OK** button.

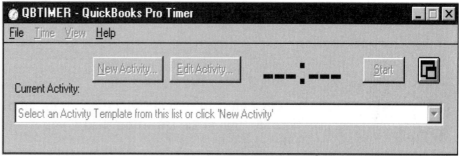

Figure 5.6 shows the QuickBooks Pro Timer© main screen.

  - b. In the **New Timer File** screen and at the **File name** field, type a name with the extension **.tdb**. This file will be created in the **qbtimer** folder and it will contain your time information in the Timer. If there is a need, you may create multiple files of this type. Click next at the **OK** button.
  - c. At the next screen, click the **NO** button.
  - d. Next, in the QuickBooks Pro Timer© select the *File* at the menu bar and the *Import QuickBooks Lists* option:
    - In the next screen, click on the **Continue** button.
    - Next, select the QuickBooks® folder, in drive **C** and the QuickBooks® folder.
    - Select the list *name* that was exported in Step A, 3 above.
  - e. Next, select the **New Activity** button (or **Time** from the menu bar, and **Create New Time Activity**) to begin with an employee's name from the **New Activity** screen:
    - Select an employee name from the **Your Name** field.
    - Select a *Customer or Job name* at the **Customer:Job** field.
    - Select a service type Item at the **Service Item** field.

- Select the **billable** box if the employee time is to be billed to a customer.

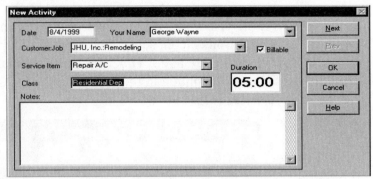

Figure 5.7 shows the New Activity screen completed.

    f. Click at the **OK** button to start keeping time for the employee's activity or you may type the time in the **Duration** field if you want to start with a time other then zero (0).

    g. In the **QuickBooks Pro Timer**© screen, there is the **Stop/Resume** button to stop and start the Timer from accumulating time.

B. In this step, you'll export the time from the QuickBooks Pro Timer© and into your company in QuickBooks Pro®:

To accomplish this task, follow these steps:

    a. Click at the **Stop** button to stop the time tracking.
    b. From the Timer's menu, select **File/Export Time Activities**
    c. In the next screen, select the **Continue** button.
    d. Set the **time** frame of the employee activity that you want to export and click **OK**.
    e. Next, enter a **name** and the extension **.iif** for the file you want to export. This file will be saved in the Timer's folder (qbtimer).

Step C:

The purpose of this step is to import the employee time from the QuickBooks Pro Timer© and into QuickBooks Pro®.

To Import time from the QuickBooks Pro Timer©, select the following:

1. The **File** from the menu bar
2. Next, select the **Timer Activities** (in QuickBooks® 2000, select **Timer**).

3. Next, select the **Import Activities from Timer** option. Next, click the **OK** button.
4. Next, select the file *name* you want to import by highlighting it, and click the **OK** button.

After you import the time into QuickBooks Pro®, you will notice on the main screen, the **QB Pro Timer Import Summary** screen. This screen, contains the **Time Activity Detail** report by Job and Employee name that you may view on the screen or print via the printer.

The same report can be obtained through the following options:
1. The **Reports** at the menu bar
2. The **Project Reports**
3. And the **Time Activity Detail** report

Note: To export through a floppy disk, follow the same steps but replace the drive, instead of C, type drive A or B depending on the configuration of your hardware.

Invoicing Billable Time

To bill a customer for **billable** time that has been tracked using the QuickBooks Pro Timer©, select the following:
1. The **Activities** at the menu bar (in QuickBooks® 2000, select **Customers**).
2. Next, select the **Create Invoice** function
3. At the **Customer:Job** field, select the *Customer* or *Job* name from the list
4. Next, select the **Time/Cost** button by clicking once on it
5. Next, in the **Choose Billable Time and Costs** screen select the **Time** tab

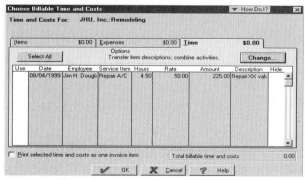

Figure 5.8 shows the **Choose Billable Time and Costs** screen.

200

6. Next, place a checkmark by clicking under the **Use** column and click at the **OK** button to create an invoice for the customer.

Export Addresses
(Mail merge)

*P*urpose: It allows you to work with names and addresses from your Customer, Vendor, Employee and Other Names from within QuickBooks® and merge them with a document in a word processor in order to create totally *customized* letters.

This process is very efficient because it allows you to create and print large quantities of letters quickly. This function can become a powerful tool to be used for marketing purposes to contact prospects or customers or simply to improve relations with employees and vendors.

To **Export Addresses** (mail merge):
1. Select **File** at the menu bar (in QuickBooks 2000, select **File** and **Utilities**).
2. The **Export Addresses** (or *Mail Merge* for older versions).
3. Next, select the *file* you want to export (merge) such as Customer, Other Names, etc.
4. Next, at the **Save Address Data File** screen, type a *name* for this new file with the extension **.txt** and click the **Save** button.
5. Next, start your **Word Processor** and:
    a. For QuickBooks® 2000 only, you need to change the data file into a *text table*. If you are using MS Word, after you open the file, highlight it and select *Table/Convert/Text Table* and save as a Word document. For all other QuickBooks® versions start your Word Processor and select **File** at the menu bar and the **Open** option to open your document you have created in the word processor or create a new document before you proceed
    b. Next, select **Tools** at the menu bar and the **Mail Merge** option
    c. At the **Mail Merge Helper** window, click at the **Create** button and select the **Form Letters** option

d. Next, click at the **Get Data** button and select the **Open Data Source** option
e. Next, select drive C at the **Look in field** and click twice at the QuickBooks® folder.
f. Next, at the **Files of Type** field select **Text Files**. Next find and select the file you exported from QuickBooks® click once on it and click at the **Open** button.
g. Next, select **Edit Main Document** by clicking on it.
h. Next, place your cursor at the position on your document where you want to start and click at the new button on the upper left corner of your document called **Insert Merge Field** and select the fields you want by clicking and releasing the mouse button.
i. After you build the merge fields, select the following:
- The **Tools** menu at the menu bar
- The **Mail Merge** again and the **Merge** button
- Next, select whether you want **All** or the **Selected** names option and
- Click at the **OK** button to proceed.

Exporting Files

𝒫urpose: To create an export file from an existing company within QuickBooks® that can be used by another QuickBooks® company.

Using this QuickBooks® capability, you can take a file such as the Chart-of-Accounts, the Customer list, the Vendor list, the Items, the Payroll Items and etc., from one company within QuickBooks® and into another company.

This process of moving, exporting and importing files, can be very useful. In case that you want to create a new company in QuickBooks®, you can import (copy) any list from an existing company into the new company.

When you export a list from one company to another, the list moves without any data. For example, if you export the chart-of-accounts from company A, the accounts will move into company B without any amounts (or balances).
The process of exporting and importing works as follows: You first need to *export* from company A and second, you must *import* the same file into company B.

A. Export a file from company A:
    a. Select **File** from the menu bar
    b. The **Export** function (in QuickBooks® 2000, select **Utilities** and **Export**).
    c. Click at the *list* you want to export
    d. Click at the **OK** button
    e. Next, type a *name* with the extension **.iif** for this new file (for example, coa.iif) and click at the **OK** button and **OK** again at the next screen.
       Note that the **iif** is the QuickBooks® interchange file extension.

B. Import the exported file into company B (first create company B in QuickBooks®):
    a. Select **File** at the menu bar
    b. Next, select the **Open Company** option to open the new company.
    c. Next, select the **File** again.
    d. Next, select the **Import** option (in QuickBooks® 2000, select **File** and **Utilities**).
    e. From the **Import** screen, select the file you want to import by clicking on it once. Click the **Open** button. Click **OK** at the next screen.

Backup & Restore

*P*urpose: To ensure that there is a second copy of your company's data, other than the one found in the hard drive, that you can use in the event that there is a hardware or software failure.

A. **Steps to Back up and Restore using floppy disks:**

   1. **Formatting** disks:
      You can format disks via DOS or Windows®.
      a. To format via DOS, insert a disk into floppy drive a: or b: and at the C:, type **Format a:** or **b:** and press *Enter*.
         Through the DOS, you can also use the *unconditional* format command. To use this command, at the C: type **Format a:** or **b: /u /f:1.44** and press *Enter*.
      b. To format via Windows® 95 or 98:
- Insert a disk into floppy drive a: or b:
- Point at the **Start** button and click with the *right* mouse button.
- Next, select the **Explore** option and click with the left button.
- Select the **3$^1/_2$ Floppy (A:)** and point and click with the right button.
- Next, select **Format** from the drop down menu and click with the left button
- Next, click on the **Start** button to start formatting.

   2. About **Back up** procedures:
      a. Back up *every day*
      b. Back up into a *different* set of disks each day (one set for each working day).
      c. Remove the *current* set from the office.

   3. **To back up** your data into a floppy disk:
      a. Insert the formatted disk into drive **a:** or **b:**
      b. Start QuickBooks® and select **File** at the menu bar
      c. Next, select the **Back Up** option
      d. Make sure that at the **Back up Company to** screen, you select drive **a:** at the **Save in** field.

        e. Click at the **Save** button to proceed with the Back up of your data.
4. **Restore** data:
    a. Insert the disk you want to restore <u>from</u>, into the floppy drive
    b. Select the **File** at the menu bar
    c. Next, select the **Restore** option
    d. At the **Restore From** screen, make sure that at the **Look in** field you select drive a:
    e. Click at the **Open** button to proceed with the restoring of the data

B. **Back up into a Tape:**

If your hardware is equipped with a Tape back up device, you **may** use it to back up your accounting data.

Tape back ups have an *advantage* over floppy disk back ups when a company's file size is such that it may take several disks to back up. A tape back up also offers the choice whether to back up just the company file or the program and company files.

To back up into a tape, follow the directions available from the manufacturer of the hardware.

Note: Remember to backup daily into a different tape. Remember also, that you must <u>remove</u> the tape that you backup from the office.

Customizing Forms

*P*urpose: To learn how to customize forms that would meet your company's needs.

In QuickBooks®, you can customize forms to quite an extensive level in order to fit a particular requirement that your company may have. You can customize the following forms:
1. Invoice
2. Credit Memo
3. Cash Sale
4. Purchase Order
5. Statement
6. Estimate

To customize an **invoice**, select the following:
1. The **Lists** at the menu bar
2. Next, select the **Templates** at the drop down menu
3. Select the form you wish to customize by highlighting. Next, either click twice on the highlighted form or select the **Templates** button at the bottom of the Templates screen and the **Edit** option.
Note: If you haven't customized an **Invoice** form before, at this step you need to duplicate an original form. Original forms are listed as "Intuit Invoice".
To **duplicate** an original form, highlight the form and select the following:
   a. The **Templates** button
   b. Next, select the **Duplicate** option from the menu, just as in figure 5.9.

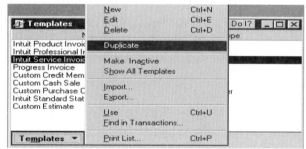

Figure 5.9 shows the form duplication template's steps.

   c. Select the form type from the next screen and click at the **OK** button as shown in figure 5.10.

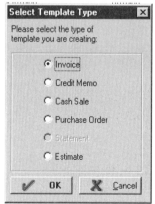

Figure 5.10 shows the form selection screen.

   d. Next, highlight the new form called "DUP": Intuit Service Form" and click at the **Templates** button.

206

e. Next, click at the **Edit** button to enter the **Customize** screen.
4. In the **Customize Invoice** screen, you will find following available: Header, Fields, Columns, Footer and Options tabs. Through this screen, you may change the following:
   a. The form's title
   b. Add or remove fields
   c. You may add or remove columns
   d. Change the column order
   e. You may change the fonts used in the form
   f. You may elect to print or not, the company name and address, depending upon whether you print on a preprinted form or on plain paper
   g. Add your company's logo and
   h. You may also move and resize objects on the form

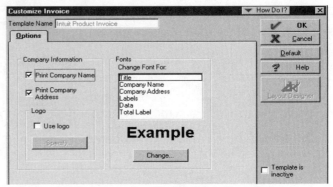

Figure 5.11 shows the **Customize Invoice** screen.

- At the **Header** tab, as in the other tabs, you may place a checkmark (√) or remove it from either the **Screen** or the **Print** columns. <u>The checkmark adds a title to the form and the absence of it removes it</u>. You can highlight a field under the **Title** column and type over a new name as in figure 5.12. All the other tabs allow you the same options.

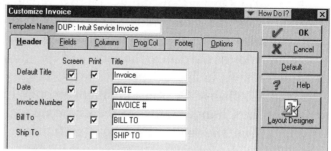

Figure 5.12 shows the **Header** tab.

- The **Column** tab, allow you to manage the columns of the form. The columns are the vertical spaces such as the **Item, Description, QTY, RATE and AMOUNT**, etc.

  As in the previous tab, you may place a checkmark (√) or remove it from either the **Screen** or the **Print** columns. The checkmark (√) adds a column to the form and the absence of it removes it. For example, on the Invoice or the Estimate form, you may choose to remove columns that show *costs* and *markups* from the printed copy that goes to the customer. All you have to do is remove the checkmark from the **Print** column.

  You can highlight a field under the **Title** column and type over a new name. The **Order** column allows you to change the order in which the columns will appear on the form, as it shows in figure 5.13

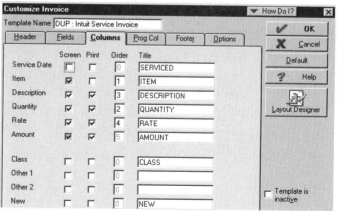

Figure 5.13 shows the **Column** tab

- The **Options** tab allows you to *add* or *remove* the company name and address from the form by clicking under the **Company Information** field. You may also change the font type, style and size of the form title, company name and address the data in the form and etc., through the **Change Font For** field as shown in figure 5.14.

To change the fonts, highlight an option, as shown in figure 5.14, and click at the **Change** button. At the **Example** screen next, you may change the font *type*, *style* and *size* as shown in figure 5.14 below.

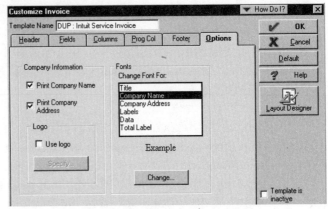

Figure 5.14 shows the **font** option screen.

To add a *logo* to the form, click at the **Use Logo** field to add a checkmark (√) as shown in figure 5.15. At the next screen, click at the **File** button and select from the available list of files, select your company's logo, by highlighting the file. Next, click at the **Open** button and the logo will be displayed at the **Selected Logo** area as in figure 5.16. Next click at the **OK** button.

Note: You first must create a logo in another program such as Paint that comes bundled with Microsoft® Windows® or other desktop publishing programs. If you do create a new logo, make sure that the new logo's file is saved in the QuickBooks® folder with the extension of .bmp added as in figure 5.16.

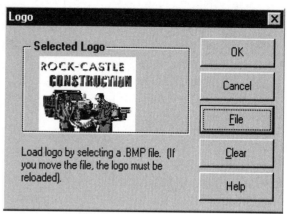

Figure 5.15 shows the **Logo** screen.

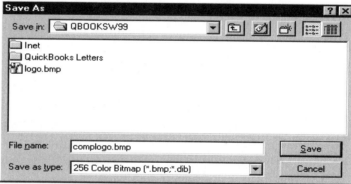

Figure 5.16 shows the new logo file in the **File name** field at the **Save As** screen.

- The **Layout Designer** button to the right of the **Customize Invoice** screen, as it shows in figure 5.11, allows you to move and resize fields. By clicking once, inside a field, as shown in figure 5.17, and holding the left mouse button, you can drag the mouse and place the now highlighted field, in a different location.
  Also, by placing your mouse cursor on a **dot** around a highlighted field (you'll see the cursor changing shape) as in figure 5.17, you can resize the field by moving the mouse to the desired position.

210

Notice also on figure 5.17 below, there is the company logo appearing at the upper left side of the invoice.

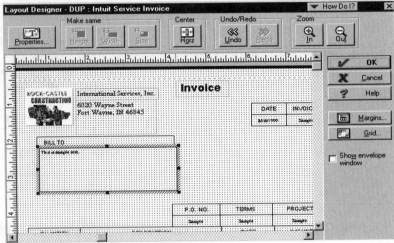

Figure 5.17 a customized Invoice template.

After you have finished with redesigning the form, click at the **OK** button and in the next step, highlight the **Template Name** field and <u>type a new name</u> for this form and you have finished redesigning a new form.

To customize other forms, follow the steps described in the Invoice example.

Budgeting

*P*urpose: Is to examine what budgets are and to enter budgeted figures in QuickBooks.

These figures then are used for comparison against the actual company performance. Account balances in the General Ledger, **are** the actual performance of the company, which is reported via the Profit and Loss Statement (report). These balances are accumulating continuously through the recording of the daily business activities such as income and expenses.

What is a budget: A budget contains *estimated* figures for the Income and the Expenses of the business. They are set by management and ought to be used as *goals* that the company as a whole, must reach during a given time period.
Budgets should be compared against the actual figures on a monthly, quarterly or yearly, regular basis.

Because budgets are simply estimated figures, they may vary *drastically* from the actual figures.

To arrive at these estimated or budgeted figures, companies use various methods. Some of these methods are quite complex such as the application of computerized statistical forecasting methods used on powerful computers. Other companies simply use plain averages taken from a prior year. Both of these methods are quite the extreme. However, a common sense, simple to use method that small business could use is to seek the knowledge of experienced employees, and feedback from customers and vendors it deals with, regarding past performances and future economic developments, plans for growth, etc. in order to arrive at reasonable figures that can be used as budgets.

After the budgets are created, they should be entered into QuickBooks®.

To enter budgets in QuickBooks:
Select *Activities/Other Activities/Set Up Budget*.
In QuickBooks 2000, select *Company* and *Set up Budgets*.

To create a budget for a Customer:Job, follow these steps:
To budget for all the income and expenses related to a customer or job, leave the Account and Class fields blank
1. In the **Set Up Budgets** window, choose the **Fiscal Year**
2. Next, select a **Customer:Job**.
3. Enter the budget amount for the first month, and then fill in the remaining months. You can use the **Fill Down** button to speed up entering the amounts.
4. Click **Save**. If you are budgeting for several customers or jobs, choose another one and repeat these steps. When you are finished, click **OK**.
   To create a budget for a project (a job) but not month by month:
   - Select a Customer:Job from your list (you do not have to use an account)
   - Enter the budget for the entire project in the <u>first month's</u> space
   - Click **OK**.

To create a budget for Income and Expense accounts follow these steps:
1. In the **Set Up Budgets** window, select the **Fiscal Year**.
2. Next, select an **Account**.
3. Enter the budget amount for the first month, then fill in the remaining months. Use the **Fill Down** button to speed up entering the amounts.
4. Click **Save**. If you are budgeting for several accounts, choose another account and repeat these steps. When you are finished, click **OK**.
Remember: To see variances between budgeted and actual amounts by time period, set up budgets for accounts without a customer, job or class.

To create a budget for a Class, follow these steps:
To create a budget that covers all expenses and income for a particular class, leave the Account and Customer:Job fields blank.
1. At the **Set Up Budgets** window, select the **Fiscal Year**
2. Next, choose a **Class**
3. Enter the budget amount for the first month, and then fill in the remaining months. Use the **Fill Down** button to speed up entering the amounts
4. Click **Save**. If you are budgeting for several classes, choose another one and repeat these steps. When you are finished, click **OK**.

To copy a budget to a new year:
1. From the **File** menu, select **Utilities** and **Export**
2. In the **Export** window, select **Budgets**

3. Click **OK**

4. In the **Export** window, enter a filename (like mybudget.iif) with the extension: **.iif**

5. Open the file mybudget.iif with a spreadsheet editor, such as the Microsoft Excel®

6. Change the year in the column called STARTDATE and **save** the changes

7. Enter the QuickBooks® program and select **File/Utilities** and **Import**

8. Select the file you named **mybudget.iif** in which you changed the date in the spreadsheet.

Reports:
To create a P/L budget vs. actual report:
From the **Reports** menu, select **Budget** and **Profit & Loss Budget vs. Actual**.
*Note: Make sure to select the **Customize** button and at the **Columns** window, select a time frame such as Total, Month, Quarter, etc. In the **Row Axis** window, select Account.*

To create a P/L by job overview report:
From the **Reports** menu, choose **Budget** and **Profit & Loss Budget by Job Overview**.
*Note: Make sure to select the **Customize** button and at the **Columns** window, select a time frame such as Total, Month, Quarter, etc. In the **Row Axis** window, select Customer:Job.*

To create a P/L budget vs. actual by job report:
From the **Reports** menu, select **Budget** and **Profit & Loss Budget vs. Actual by Job**.
*Note: Make sure to select the **Customize** button and at the **Columns** window, select Account. In the **Row Axis** window, select Customer:Job.*

Year-end Process

At the end of your company's fiscal year, there is certain activity that <u>must</u> take place. Part of this activity, is taken care of automatically by QuickBooks®. The other part, you must play an active role to complete.

When you reach the end of your company's fiscal year, QuickBooks® automatically transfers the bottom line which the *Net Income* or the *Loss* from the P&L statement into the **Retained Earnings** account and "empties" the P&L by placing zeros on all the *Income* and the *Expense* accounts. The zeroing of the income and the expense accounts, allows you to start the new fiscal year with a brand new Profit and Loss statement.
Note: For an explanation of the <u>Retained Earnings</u> account, please see the glossary section below.

At the end of the fiscal year, you must be prepared to record certain transactions that would complete the year, and allow you to print *accurate* year-end financial statements and tax information.
The following **Year-end transactions** may apply to your company's situation:
1. Record the *depreciation* transaction
2. Record *interest* earned from interest bearing accounts at the bank or elsewhere
3. Record *bank fees* deducted from your account at the bank
4. Record the adjustment for the *prepayment* of various expenses
5. Record the adjustment to *loan* (liability) accounts according to bank records for any interest and principle differences that often exist
6. Record any adjustments that may be necessary accounts with incorrect balances in order to bring their balances to the proper amounts.
Please note: Consult with your accountant before you adjust any account balances.

QuickBooks® Keyboard Shortcuts

Purpose	Key	Purpose	Key
Cancel	Esc	Display Info about QuickBooks®	Ctrl +1
Display Info about QuickBooks®	Ctrl +1	Move to next field	Tab
Edit transaction selected in register	Ctrl + E	Move to previous field	Shift –Tab
Delete line from detail area	Ctrl + Del	Show chart-of-accounts	Ctrl + A
Insert line in detail area	Ctrl + Ins	To start Write Checks	Ctrl + W
Cut selected characters	Ctrl + X	To start Create Invoices	Ctrl + I
Copy selected characters	Ctrl + C	Find transaction	Ctrl + F
Paste cut or copied characters	Ctrl + V	Quick report on list items	Ctrl + Q

Glossary

1099 Form: The federal form used for the annual reporting of payments made to non-employees.

940 Form: The federal form used for reporting of the Federal Unemployment Tax (FUTA).

941 Form: The federal form used for reporting the *quarterly* wages and the FICA taxes withheld from employees' paychecks.

Accrual Basis Accounting: Is the type of accounting that allows for *income* and *expenses* to be recorded in the period (month) they have accrued.

Alphanumeric: A word containing alphabetic and numeric characters (numbers and letters).

Amortize: To Liquidate on an installment basis[1].

An Entry: A part of a business transaction. A business transaction always has *two* or more entries.

Book Value: The accounting value of an asset.

Break Even (or Break Even Point): The point of the volume in sales at which the total *revenues* equal total *expenses* and therefore, profit is zero[1].

Business Risk: The basic risk inherent in a company's operations. The risk a person assumes at the time of starting a business.

Capital Asset: An asset with a life of more than one year that is not bought and sold in the ordinary course of business[1].

Capital Expenditure: The acquisition cost of a business asset. This expenditure is debited directly to an appropriate asset account[1].

Cash Basis Accounting: Is the type of accounting that allows recording of *income* and *expenses* at the time of payment.

Cash Flow: A summary of the *sources* and *uses* of funds for a company over a period of time.

Chart of Accounts: A listing of all the accounts used in a company through which all the business transactions are recorded.

Default: The starting position of a *character*. A *word* or a function that the comes up with when you enter into a particular part of QuickBooks®.

Depreciation: The process of recording the reduction of an asset's value due to *wear*.

Double Entry: The method of accounting used to record a business's activities and it requires two or more entries for every single transaction.

FICA (Federal Insurance Contributions Act): The amount of Social Security and Medicare withheld from an employee's paycheck and is matched by the employer.

Fiscal Year: The twelve-month accounting period used by companies to record a business's activity.

Interest: The associated cost with borrowing money. The cost of money

Post: The process that completes the recording of a transaction when you click the **OK** or the **Next** button in the QuickBooks® software program. According to manual accounting terms, Posting is the process where transactions physically are moved from the General Journal to the General Ledger.

Profit Margin: The ratio of profits after taxes to total sales[1].

Retained Earnings: It is an Equity type account in the chart-of-accounts. Its purpose is to contain the income (or loss) the company realizes at the end of each fiscal year. The figure in the Retained Earnings account is the sum of <u>all</u> the earnings throughout the company's history.

Transaction: A business *activity* such as a purchase, a payment, a customer invoices, etc.

[1] Source: Managerial Finance By J. Fred Weston and Eugene F. Bringham

Chapter 6

Reporting

$\mathcal{T}$he purpose of this chapter is to help you understand the importance and how to use the Profit and Loss statement, the Balance Sheet, the Trial Balance and other important reports.

In this chapter we'll examine the following QuickBooks® reports:

- **Profit & Loss Statement**
- **Balance Sheet**
- **Trial Balance**
- **A/P Report**
- **A/R Report**
- **Deposit Report**
- **Job Profitability Report**

Profit & Loss Statement

The Profit & Loss Statement (P/L) shows you the *operating performance* of your company over a specific period of time such as a month or on a cumulative basis, such as year-to-date.
The Profit and Loss statement (P&L) it's a report you should print **every week**.

It is also important to print and save the month-end P&L report into a separate folder.

The P/L report is also known as the Income Statement.

To print the P/L:
1. Select **Reports** at the menu bar (in QuickBooks 2000, select **Reports** and **Company & Financials**).
2. The **Profit & Loss** option that you want to see.
3. Next, select the **Standard** option for the current month's P/L, or the *YTD Comparison* option for a comparison of the current month's data vs. year-to-date.
   (In QuickBooks 2000, select **Profit & Loss Standard**).

The manager must analyze the P/L weekly. During the analysis, the manager should pay close attention to the <u>sales</u> and the <u>expense</u> accounts and understand the detail activity of the purchases and the sales figures for every weekly and monthly period.
In the process of analyzing the company's performance, the actual figures should be compared against budgeted or planed figures. The budget or plan, should contain a list of up to five objectives among which there ought to be two that pertain to the P/L:
1. One objective should be the amount of total **sales** (or gross sales) the company should achieve each year. This objective should be broken down into monthly figures.
2. The second objective is the **income** the company ought to make from the sale of its product or service.

The amount of *net income* shown on the P&L indicates the level of *profitability* the company is achieving. When the company is profitable, it indicates that it is *efficient*.

On a monthly basis there should be a **trend** analysis of the P/L where all the income and expenses accounts are compared against a *budget* and a *prior year* performance.

A good format for the P/L is the one that shows three columns: The current month, year-to-date and the percent columns.

To **format** your P/L into three columns, select the following:
1. Follow steps 1-3 above and
2. Once in the report, select the **Customize** button on the upper left corner of the screen.

Note: the percent column applies to the current month column.

International Services, Inc.
Profit and Loss

	Jun '99	Jan - Jun '99	% of Income
Ordinary Income/Expense			
Income			
4050 · Sales - Parts	9,710.00	215,210.00	41.3%
4070 · Sales - Service Fees	14,600.00	203,600.00	62%
4080 · Sales - Other	59.00	59.00	0.3%
4090 · Customer Discounts	-829.85	-829.85	-3.5%
Total Income	23,539.15	418,039.15	100.0%
Cost of Goods Sold			
5000 · Cost of material	13,491.23	173,491.23	57.3%
Total COGS	13,491.23	173,491.23	57.3%
Gross Profit	10,047.92	244,547.92	42.7%
Expense			

Figure 6.1 shows a portion of a three-column **P/L** statement.

Balance Sheet

*T*his report shows the *financial position* of the company as of the specific time it is printed. The financial position is stated through the amounts found in the company's Asset, Liability and Equity accounts.

It is recommended that you print the Balance Sheet **every week.**

To print the **Balance Sheet**:
1. Select the **Reports** at the menu bar (in QuickBooks 2000, select **Reports** and **Company & Financials**).
2. The **Balance Sheet** option.

3. Select the **Standard** option or the **Comparison**. The *Standard* shows the current year and *Comparison* shows the current year vs. last year.

The manager on a weekly basis, should examine the entire report but pay close attention to the following accounts:
1. The amount of cash available in the *Bank account (s)* and the account's activity by transaction.
2. The balance in the *A/R account*. Make sure amounts owed to the company by customers are received within the time due. The A/R account plays a key role in the company's **cash flow**.
3. The balance in *A/P account*. Remember to print the *A/P Aging Detail* report for the detail of the account.
4. The balances in the *payroll liability accounts*.
   Note: Remember to pay payroll taxes due to Federal and State governments on time.

International Services, Inc.
Balance Sheet
As of June 30, 1999

	Jun 30, '99
ASSETS	
Current Assets	
Checking/Savings	
1000 · Checking Account- NBD Bank	13,952.69
1010 · Checking Account- HJK Bank	12,405.96
1022 · Checkind Acc.	494.90
Total Checking/Savings	26,853.55
Accounts Receivable	
1200 · Accounts Receivable	13,389.00
Total Accounts Receivable	13,389.00
Other Current Assets	
1113 · Prepaid Expenses	11,000.00
Advances	

Figure 6.2 shows a portion of a Standard type **Balance Sheet.**

Trial Balance

This report shows all the General Ledger account balances on a cumulative basis as of a given time period. You should consider printing this report on a monthly basis and save each copy into a separate folder together with a month-end P/L and Balance Sheet.

To print the Trial Balance report:
1. Select the **Reports** at the menu bar
2. Next, select **Other Reports** (in QuickBooks 2000, select **Reports** and **Accountant & Taxes**).
3. Next, select the **Trial Balance** option.

International Services, Inc.
Trial Balance
As of June 30, 1999

	Jun 30, '99	
	Debit	Credit
1000 · Checking Account- NBD Bank	13,952.69	
1010 · Checking Account- HJK Bank	12,405.96	
1022 · Checkind Acc.	494.90	
1200 · Accounts Receivable	13,389.00	
1113 · Prepaid Expenses	11,000.00	
Advances:1510 · Employee Advances	1,320.00	
Advances:1520 · Loan / Mike N.	0.00	
1020 · Petty Cash	444.00	
1030 · Savings Account	45.00	
1035 · Money Market Funds	2,500.00	
1120 · Inventory	12,129.70	
1499 · Undeposited Funds	271.88	

Figure 6.3 shows a portion of a **Trial Balance** report.

A/P Report

The purchases you record in QuickBooks® via the Vendor bills, they become part of the **Accounts Payable** (A/P) report whose purpose is to provide you with detail information of the various amounts your company owes to vendors and suppliers. The people or companies you do business with by purchasing their product and services.

There is also the **Accounts Payable** account, which is one of the accounts in the chart-of-

accounts. This account is reported in the *current liability* section of the **Balance Sheet**.
To print the **A/P report**, select the following:
1. The **Reports** at the menu bar.
2. At the **A/P Reports** option below (QuickBooks 2000, select **Reports** and **Vendor & Payables**).
3. Select either the **A/P Aging Summary** or the **Aging Detail** report.

The *Detail* is the most preferred report because it shows every transaction separately. It is one of the reports you need to print on a *weekly* basis alone with the P/L and the Balance sheet.

A/R Report

This report contains all the amounts owed to your company by customers.

When you create an invoice for a customer, it automatically goes into the **Accounts Receivable** (A/R) report. This report should be printed on a weekly basis.

There is also the **Accounts Receivable** account, which is one of the accounts in the chart-of-accounts. This account is reported in the *current asset* section of the **Balance Sheet**.

Business tip: The A/R report plays an important role in the cash flow of your business and it should be printed on a weekly basis. It will show you the amounts due and the age of each amount.
To print the **A/R** reports select:
1. The **Reports** at the menu bar
2. The **A/R Reports** (in QuickBooks 2000, select **Reports** and **Customers & Receivables**).
3. Next, select the **A/R Aging Detail** or the **Summary** report

Deposit Report

The **deposit** report is an important report you may print every time you make a deposit. When you print this report, you will have a record of all the deposits by *name, amount,* and *check* number you've received.
To print the **Deposit** report, select the following:
1. The **Reports** at the menu bar

2. The **Other Reports** at the drop down menu (QuickBooks 2000, select **Reports** and **Banking**)
4. And the **Deposit Detail** report

Job Profitability Report

It's a very valuable report that you should print for *each* job your company is currently working. This report should be printed on a *weekly* basis and the manager must spend time analyzing each expense transaction that appears in a particular job's report.

Every job that your company spends time and utilizes its valuable resources on <u>must</u> be profitable. If it is not, you as the manager must find the reason and be prepared to <u>make</u> changes.
To be able to make good decisions as a manager, you must have *accurate* information. And accurate information comes from recording transactions by using proper accounting.

To print the **job** cost report, select the following:
1. The **Reports** at the menu bar
2. Next, the **Project Reports** (in QuickBooks 2000, select **Reports** and **Jobs & Time**).
3. Next, select the **Job Profitability Summary or Detail**.
   The **Job Profitability** report is a *profit* and *loss* report. It shows you how the company is performing on a per job basis. The detail report shows all the transactions regarding income, expenses (or costs) and the profit (or the loss) the company is making.
   To print a report by **job type**, select the following:
   a. The Filters button at the top of the report window
   b. Under the Filter column, select the **Job Type** option by scrolling
   c. At the **Job Type** window to the right, select a particular job type name or the selected Job Types option by clicking. Next, click **OK** to print the report.

Getting the Lists

For your convenience, we have placed the *Chart-of-Accounts*, *Items*, and *Payroll Items* lists that came with the book, on our **web site so** that you may access them immediately without the need of any further work.

To *download* this information, simply follow the steps below:

1. Optionally, create a new folder in your drive **(C:)** and name it **Download.**
2. Start your Internet browser.
4. Type our website address: www.systemmanagement.com
5. Once you are in our website:
    a. Click at the **Download** button (below).
    b. At the next screen, fill out the **form** completely.
    c. Click at the **Submit form** button. Next, click at the **LOGIN** button.
    d. Next, type the following user ID: **qbbooks**. Password: **gd1**
    e. Next, **right click** on the file you want to download
    f. At the "**Look in**" field on the menu, select drive **(C:)**. **Next**, select the **folder** you've created (Download) by clicking twice on it.
    g. Click the **Save** button.
    h. To import the **List** (s) into your company file, in the QuickBooks® software, please follow the instructions on page # 26.

INDEX

940 Form, 179, 217
941 Form, 178, 217

A/P report, 91, 224
A/P Report, 219, 224
A/R Report, 126, 219, 224
Account field, 21, 23, 39, 40, 43, 72, 98, 103, 106, 108, 121, 151, 183, 188
Accountant's Review, viii, 182, 194, 195
Accounts Affected button, 21, 23
Activities menu, 3
Activity Report, 29, 31
Adding an employee, 16
Addition items, 14
Address Info tab, 17
Adjust QTY/Value on Hand, 72
Affect liabilities and bank account option, 23
Asset category, 24
Audit Trail, 75, 76
Automatically Enter, 64, 66

Back Up & Restore, 182, 204
Bad Debt, 44, 133
Balance Sheet, viii, 24, 71, 83, 86, 92, 96, 97, 219, 221, 222, 223, 224
Bank Account Reconciliation, 182, 188
Billable Expenses, 56

C.O.D, 48, 51, 54
Charge, 94, 95, 96, 188
chart-of-accounts, 7
Chart-of-Accounts, 1, 9, 24, 27, 67, 105
Checking account, 44, 48, 49, 53, 62, 71, 77, 101, 106, 108, 110, 118, 132, 144, 146, 175, 186, 187, 188, 189
checks, 3, 5, 27, 35, 42, 51, 60, 68, 102, 105, 108, 109, 117, 131, 143, 144, 148, 158, 161, 163, 164, 173, 176, 177, 192

Class, 1, 17, 45, 46, 54, 58, 59, 62, 68, 69, 72, 80, 84, 94, 95, 113, 124, 137, 166, 168, 170, 175
Class field, 17, 45, 58, 59, 68, 107, 113, 166, 168, 170
Classes, 8
Columns tab, 115
Company Preferences, 26
Compensation screen, 12
Converting P.O's into Bills, 87
Create Credit Memos/Refunds, 48, 112, 134, 136
Create Estimates, 45, 112, 113
Create Invoices, 35, 45, 49, 57, 112, 124, 129, 216
Create Statements, 112, 142, 143
Credit Card balances, 11
Credit Card Charges, 45, 48, 50, 55, 92, 94, 95, 99
current date, 5
Custom Reports, 121
Customer, 3, 4, 5, 10, 30, 31, 32, 42, 43, 49, 54, 55, 56, 57, 58, 62, 63, 68, 69, 72, 80, 84, 85, 86, 94, 95, 112, 113, 116, 117, 121, 129, 130, 133, 135, 137, 140, 143, 144, 168, 170, 172, 174, 175, 192, 197, 198, 200, 201
Customer:Job, 30
Customizing Forms, 182, 205

Deduction type, 13, 154, 156, 158
Deposit Report, 123, 219, 225
Discount, 38, 43, 47, 103, 104, 113

Easy Step Interview, 6
Edit menu, 3, 4
Edit Paychecks, 175
Editing, Voiding and Deleting, 73
Employee *name*, 5, 68, 200
Employee YTD summaries, 20

Ending Balance, 98, 99, 188
Enter Bills, 35, 45, 48, 49, 55, 83, 84, 85, 86, 87, 101
Enter Cash Sales, 35, 46, 55, 112, 114, 116, 118, 119, 131, 136, 137, 138
Enter Employees, 147, 165
Enter Statement Charges, 112, 140
Enter Vendor, 50
Entering Beginning Balances, 1, 29, 31, 34
Entering Customers, 1, 30, 55, 85
Equity category, 24
Expense category, 24
Export Addresses, 182, 201
Exporting Files, 203

Federal ID, 5
Federal taxes screen, 15
File menu, 3, 4
Form 1065, 6
Form 1099, 110
Form 1120, 6
Form 1120S, 6
FUTA, 11, 12, 162, 166, 179, 217

Glossary, 8, 182, 217
Group, 38, 117, 118, 119, 130, 131, 137, 158, 159, 160

Hourly type, 13

iconbar, 2
Import option, 27, 203
Income & Expense tab, 9
Income category, 24
information you need, 4
Inventory Part, 9, 37, 38, 40, 41, 52, 70, 72, 88, 113, 125, 126, 135, 138
Items, 1, 3, 4, 9, 11, 13, 14, 15, 17, 19, 21, 26, 27, 37, 38, 39, 40, 52, 53, 57, 70, 80, 87, 88, 96, 113, 114, 124, 127, 135, 138, 144, 147, 148, 150, 151, 152, 154, 155, 156, 157, 159, 161, 162, 163, 164, 165, 166, 167, 203

Job Costing, viii, 10, 54, 55, 107

Job Profitability Report, 33, 219, 225
Job Types, 33, 225
Journal Entries, 29, 31, 35, 45, 48, 86, 93, 145, 146, 182, 183, 185, 187

Keyboard Shortcuts, 182, 216

Layout Designer, 210
Liabilities category, 24
Line of Credit open balance, 11
Lists, 3, 4, 9, 11, 16, 25, 27, 28, 29, 30, 32, 33, 38, 46, 47, 65, 74, 77, 81, 90, 91, 105, 111, 114, 150, 161, 163, 164, 165, 175, 182, 197, 198, 206
Long Term Liability, 186
LT Loan, 185

Mail merge, 201
Make Deposits, 48, 49, 112, 118, 119, 120, 122, 131, 132
manual check, 70, 106, 190
Memorizing Transactions, 63
menu bar, 2

Navigator, 2
Non-inventory, 37
Note payable balance, 11

Opening Balance, 25, 28, 29, 31, 92, 93, 101, 191
Opening Balances, 9, 10
Other Charge, 37, 42, 44, 45, 117, 144
Other Income type, 45, 110, 141, 142, 145
Other Reports option, 76, 123

Pay Employees, 36, 147, 173
Pay Sales Tax, 49, 50, 108, 109
Pay the Liabilities and Taxes, 147, 176
Pay Vendor, 50
Payment, 22, 38, 100, 102, 106, 117, 130, 131, 132
Payroll, viii, 3, 4, 5, 11, 13, 14, 15, 17, 19, 20, 21, 22, 24, 26, 27, 35, 36, 49, 59, 68, 138, 147, 148, 150, 152, 154, 155, 156, 157, 159, 160,

161, 162, 163, 164, 165, 166, 167, 168, 170, 173, 176, 178, 180, 181, 197, 203
Payroll Info tab, 17, 165, 166
Petty Cash, 24, 50, 76, 77, 78, 122
Preferences, 8, 26, 65, 75, 111, 128, 129, 141, 161
Prepaid Expenses, 185
Prepayments, 37, 42, 43, 112, 116
Print Paychecks, 175
Printing Checks, 59
Prior Payments, 23
Profit & Loss Statement, 219, 220
Progress Billing, 113, 128, 129
Progress Invoicing, 8
Project Reports, 55, 171, 200, 225
Purchase Discounts, 102
Purchase Order, 80, 81, 82, 83, 87, 88, 89, 90, 205
Purchase Orders, 50, 80, 81, 82, 83, 89, 90, 91
purchases, 3, 5, 28, 35, 48, 50, 63, 71, 76, 77, 80, 83, 85, 86, 92, 94, 220, 224
Purchases & Payments, 50

Qty on Hand, 41
Quick Add, 52
QuickReport, 29, 31, 32

Rate field, 12, 39, 166
Receive Payments, 49, 112, 117, 118, 119, 130, 134, 135, 136, 143
Reconciling, 97
Register, 27, 48, 50, 53, 74, 75, 77, 78, 101, 105, 106, 107, 175
Reimburse an employee, 68
Remind Me type, 65
Reports menu, 3, 4

Salary type, 13
sales, 24, 35, 38, 40, 41, 42, 49, 59, 108, 109, 110, 138, 217, 220
Sales Tax Credit, 109
Sales Tax Group, 38
Sales Tax Item, 38, 42
Save As screen, 7

Service, 9, 37, 38, 39, 40, 44, 81, 113, 126, 133, 135, 136, 137, 138, 139, 153, 168, 170, 172, 176, 179, 188, 198, 207
Set Up YTD Amounts, 20, 36
Shortcut icon, 2
Sick/Vacation button, 19, 167
Single Activity, 168, 169
Start, 2
Start button, 2, 169, 204
start date, 5
Starting QuickBooks®, 2
State taxes screen, 15
State Withholding, 12, 164
Subtotal, 38, 42, 113
SUTA, 11, 162

Taxable field, 39
Templates screen, 115, 206
Terms, 1, 28, 38, 47, 80, 124
The Accounting Quickhelp, 48
The Chart-of-Accounts, 3, 4, 8, 24, 26, 74, 77, 92, 105, 175
The **Edit** menu, 3, 4
The **File** menu, 26
The iconbar, 3
the **Make Journal Entry**, 26, 110, 183, 185
The menu bar, 3
The Navigator, 2
The Register, 50
The **Reports** menu, 3, 4
Time Reports, 171
Time tracking, 8, 174
Timer Activities, viii, 182, 197, 199
To do **Job Costing**, 55
Track Time, 147
Transfer Funds, 50, 79
Trial Balance, 5, 35, 36, 195, 196, 219, 223

Unbilled Costs by Job, 57
undeposited funds, 117, 118, 119, 130, 131, 137
Using Classes, 58
Using the Disk, 26

Vendor, 3, 4, 5, 10, 28, 29, 39, 40, 50, 52, 63, 77, 80, 83, 84, 87, 106, 107, 111, 192, 197, 201, 224

W-2 Form, 181
Weekly Timesheet, 168, 169, 170

Write Checks, 35, 45, 48, 49, 50, 51, 52, 55, 61, 66, 68, 73, 76, 100, 135, 185, 190, 216
Write Letters, 182, 192

Year-end Activity, 182
Year-end transactions, 215

231